D1322991

Teaching Physics

to KS4

Ann **Jerram**

non-specialist **handbook**

Series Editor: Elaine **Wilson**

One Week Loan

...ughton

...ADLINE GROUP

We are grateful to the NEAB for permission to reproduce examination questions. The answers given within the book are entirely the responsibility of the author and have not been provided or approved by NEAB.

Orders: please contact Bookpoint Ltd, 130 Milton Park, Abingdon, Oxon OX14 4SB. Telephone: (44) 01235 827720, Fax: (44) 01235 400454. Lines are open from 9.00–6.00, Monday to Saturday, with a 24-hour message answering service.

You can also order through our website www.hodderheadline.co.uk

A catalogue record for this title is available from The British Library

ISBN 0 340 75340 4

First published 1999
Impression number   10   9   8   7   6   5
Year   2005   2004

Cover photo from Science Photo Library
Typeset by Wearset, Boldon , Tyne and Wear.
Printed in Great Britain for Hodder & Stoughton Educational, a division of Hodder Headline, 338 Euston Road, London NW1 3BH by CPI Bath

Preface

To all the teachers and pupils from whom I have learned
and continue to learn about education and science.

My grateful thanks to Ian for so much encouragement and help.

Contents

v

6

Radioactivity 189

1 Electricity and Magnetism

chapter 1

Static charge

BACKGROUND

Electrostatic forces hold atoms together but in general are not as apparent in everyday life as gravitational forces. At one time or another you will probably have rubbed an inflated balloon on a woolly jumper and then placed the balloon against the ceiling of a dry room. Because the electrostatic force on the balloon and ceiling is sufficiently large that it exceeds the gravitational force on the balloon, the balloon sticks to the ceiling.

Clearly bodies do not always experience these electrostatic forces. The idea of electric charge, called electrostatic charge, was investigated extensively in the nineteenth century by Charles Augustin de Coulomb, who deduced the form of the law of electrostatic force. This essentially suggests that:

1 You can not tell simply by looking at an object whether it is charged or not. The only way to be sure is to see whether it can produce an electrostatic force.

2 There are only two types of charge. The property of charge cancellation has led to the two types of charge being called positive ($+$) and negative ($-$).
  - Bodies carrying the *same* type of charge *repel* each other.
  - Bodies carrying *different* types of charge *attract* one another.
  - Bodies carrying one type of charge become electrically neutral, and do not exert any electrostatic forces, after absorbing an equal quantity of the other charge type.

3 All matter contains electric charge.
- Electrons carry negative charges.
- Protons carry positive charges.
- In electrically neutral material there are equal numbers of electrons and protons.

4 When different materials are brought into close contact and energy is provided, perhaps by rubbing, electrons may be transferred from one material to the other. The direction of transfer depends on the properties of the material.
- A plastic ruler rubbed with a woollen cloth acquires electrons.
- A glass rod rubbed with a piece of silk loses electrons.

5 In the rubbing process no charge has been created, instead existing charges have been separated and redistributed.

Students find the practical work associated with this interesting and will all have their own examples of receiving, or giving, shocks. Separating charges by friction is much easier to achieve in non-conducting materials, so plastics and other synthetics make this a familiar occurrence. However, students can easily become muddled trying to use the concept of charge to explain these effects, so they need very careful guidance. You will need to be consistent in describing which charges move, and under what circumstances, to ease the understanding of the flow of charge, that is, electric current.

KEY STAGE 3 CONCEPTS

Pupils should be taught that:

- An *insulating material can be charged by friction*. There are lots of attention-grabbing experiments here; the important idea to get across in the first stage is that 'charging' means separating charges in such a way that the negative ones move and the positive ones remain. It is not necessary to refer to electrons now. Accompany each effect with a diagram showing how charges are positioned so pupils build up a clear picture in their minds. They should try to charge up different materials and, if ready, can be given ideas about earthing, that is, charge leaking away and being lost in the earth (adding a few electrical charges to all those in the earth is like adding a drop of water to the ocean).
- *There are forces of attraction between positive and negative charges, and forces of repulsion between like charges.* The rule for this interaction is *like charges repel, unlike charges attract.* These are odd uses of the words 'like' and 'unlike'.

Students who find difficulty can have simple substitutions: **like = the same** and **repel = push away**. Avoid personalising charges, such as 'these charges like each other', as it can lead to more confusion later.

A charged object such as a comb, brought near to an uncharged object such as tissue paper, can cause opposite charges in the paper to be pulled towards the comb and similar charges to be repelled. This means that the comb attracts the nearest side of the paper and repels the furthest side. These charges are said to be induced.

Some pupils will link this last rule with magnetism. It is useful to mention the similarities but also to ensure that pupils do not confuse poles and charges. A paragraph of writing about the two effects for pupils to complete the gaps will help to clarify.

The electrostatic charge generator (Van der Graaff) machine is a lot of fun and although students will not understand all the possible demonstrations at this stage, they should be allowed to see them because of the familiarity and anticipation they will experience when they revisit in Key Stage 4. Charges on the rubber belt are separated by friction at the lower roller and then travel to the top on the belt. A row of points collects the charge at the top and transmits it to the dome. Since they all repel one another the charge spreads out over the outside of the dome (see Figure 1.1). If your generator is the motorised variety ensure that pupils don't think you are feeding in electrical charges from the mains supply.

It is good practice to link back the work with that done on forces so that students may be able to suggest how and in what units such forces might be measured. They should consider where the energy comes from to separate the charges.

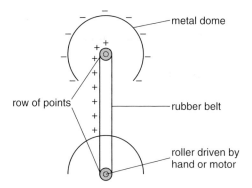

Figure 1.1 *A Van der Graaf generator. The type of charge depends on the properties of the roller and belt.*

KEY STAGE 3 ACTIVITIES

Electrostatic effects

Simple investigations include:

1 Sticking rubbed balloons to the wall – rubbing on an acrylic sweater works well.

2 Picking up pieces of tissue with a rubbed comb or plastic rod.

3 Hair-raising with a balloon near fine, dry, recently washed hair.

4 Strips of clingfilm newly off the roll or rubbed are attracted to your hand.

5 Bending a stream of water (hold a charged plastic rod near to a steady stream of water and the water will bend towards the rod).

Pieces of foil should also be tried rather like the clingfilm.

Discuss the findings with the class, moving towards the explanation as follows.

- *How can we explain what is happening here?* Discuss the electrical charges on the plastic objects.
- *What causes them?* The rubbing – explain that we call it *charging*.
- *Why do the balloons fall down eventually?* The charge leaks away and mixes up with others in the air.
- *Where were the charges to begin with?* There are two sorts of charge, called positive and negative. All materials contain them. We picture them with positive and negative charges balanced.

When we use friction the **negative** charges can be transferred from one material to another e.g. from synthetic clothing to polythene. This leaves unbalanced positive charges behind. Explain that we only know by investigation which charges occur on which materials. Students don't have to memorise these.

The individual effects are explained as follows:

1 The balloon becomes charged. Placed near the wall it induces the opposite charges in the surface of the wall and the force between them keeps the balloon in place. Eventually charge leaks through the atmosphere, particularly if it is damp, and the balloon falls.

2 The rod becomes charged, though we don't know at this stage whether it is positive or negative. The part of the paper nearest to the rod has an opposite charge induced and is attracted to the rod.

3 Induced charges in the hair.

4 The clingfilm is electrostatic due to friction in manufacture and induces opposite charges on your hand.

5 Induced charges in the water.

The foil does not display these effects because it conducts; any charges created by friction are able to move away instead of staying in place as on an insulator.

Students next investigate the interaction between charged objects. The two strips of clingfilm have already shown repulsion. They both had the same charge. Some text book methods show suspended rods balanced precariously on non-conducting threads or upturned bowls but generally pupils find this fiddly to set up. A larger scale demonstration uses two balloons, hanging on nylon thread, which are rubbed with a cloth. When using a method like this be sure to keep suspended balloons etc. away from a metal stand, try borrowing a wooden stand or use volunteers to hold them. It is clear that two similar materials charged in the same way will have the same charge. Try rods of different materials with the balloons. If one is brought near to a different charged material and is found to attract it may have the opposite charge *or* it may be neutral. Ensure pupils understand that only repulsion provides evidence for a certain charge.

The electrostatic charge generator

This is always a crowd-puller as pupils will have heard about it already. It can be disappointing in damp weather or even after several classes have been breathing heavily in an unventilated room so, if you have the opportunity, air the room beforehand and give the machine a thorough 'blow-dry' with a hair-dryer. Show pupils what is inside the dome and explain how it works (see Concepts). Switch on or turn the handle steadily, depending on whether you have a luxury or standard machine, with the small earthed ball placed a few centimetres from the larger. You can expect to see a spark of at least 3 cm between the two. Turn the machine to different angles so everyone can see. It is more effective if the room is not brightly lit. The following demonstrations show different aspects of electrostatic charge phenomena. Always have an earthed lead available so that you can discharge the dome between demonstrations. Select a few pupils and ask 'how can we explain this?'

- **head of hair** – there is usually one provided and pupils will enjoy watching as the strands all repel one another. If you want to use a pupil select someone with fine, dry, newly washed hair and make sure they stand on a rubber mat.
- **windmill** – charge streaming from the points forces the windmill round.
- **jumping balls** – these can be in a cylinder with a metal top. When the machine

provides charge the metallised balls become charged and jump up. If you touch the top with your finger or an earthed lead, the balls hit it and lose their charge and jump back down. Rapid up and down movement is seen and some pupils might realise this means that a small electric current is going through you or the lead.

- **neon lights** – a small bulb is often provided and will light up red if the leads are pointed at the charged sphere. However, a standard fluorescent tube is more 'Star Wars', especially if you hold it part of the way up. Note the fluorescent tube doesn't contain neon.

- **pupil interaction** – this can cause great excitement so is probably best left as a treat until the last five minutes and only offered to volunteers. The first pupil has one hand placed on the dome. As successive pupils link hands the last joined will feel a shock. A long line is possible if you have plenty of rubber mats or pupils have thick insulating soles.

There is no need to feel that all the effects should be understood here but just enjoyed.

Assessing pupils' learning

- Give the 'bending water' activity to do at home and describe/draw the observations and explanations. Explain why sometimes your hair flies towards to the comb when you use it.
- Design a way to measure an electrostatic force in newtons. You might consider a charged item on a newton balance (insulated from the plate) or pushing a suspended strip through an angle and measuring/calibrating the force needed.
- Ask pupils to discuss the energy changes involved and where the energy comes from to separate charges.
- Set a paragraph about how lightning occurs, leaving out words for pupils to insert from a list (**N-pole, S-pole, positive charge, negative charge, neutral**). Other contexts for the passage could be toys or interesting uses, e.g. photocopiers.
- Suggest how you could find out what sort of charge is on a new material.

ENRICHMENT AND EXTENSION ACTIVITIES

1 Find out about thunder and lightning.

2 Design a lightning conductor after discussion (important factors are a sharp point at the top and a low resistance path to earth at the base).

3 Make a toy hair-raising character using material, foil and thread which they could try on the electrostatic charge generator.

4 Collect ideas from as many people as possible for reducing shocks you receive when getting out of a car and discuss in class.

5 Explain why people are more likely to get shocks from static charges in dry weather.

KEY STAGE 4 CONCEPTS

Pupils should be taught:

- *About common electrostatic phenomena, in terms of the movement of electrons.*
  Consult schemes of work and test pupils' current knowledge about electrons. A useful model at this stage would be the Rutherford solar system picture of atomic structure with a central positive charge and negatively charged electrons in distinct orbits. The electrons can be moved from one material to another but the positive charges held by the much more massive nuclei are not able to move unless liberated, see electrolysis in the fourth concept. An everyday example is provided by the shocks experienced when stepping from a car after driving. The friction with the air as the car travels causes electrons to be transferred, leaving the car with a net charge. This charge stays on the car in dry weather since the tyres are insulating. It discharges to earth through you when you step out of the car. In damp conditions the charge leaks away through the atmosphere.

- *The dangers and uses of electrostatic charges generated in everyday situations.*
  Dangers can be demonstrated by the use of the electrostatic charge generator. Students can feel the effect of small shocks and relate them to those from clothes, nylon carpets, cars, television screens etc. Distinction should be made between these high voltage shocks, where the amount of electric charge flowing is small, and comparatively low voltage mains electric shocks where a large quantity of charge flows. A large quantity of charge is much more harmful to living tissue and students can never have too many reminders about applying safety rules at all times when dealing with electricity.

  The breakdown potential for air, about 30 000 volts per centimetre, allows comparison between the sparks seen from the charge generator and lightning. The main danger is that of the spark lighting inflammable material. Thus the discharge generated when driving, described above, could ignite petrol vapour. The hazard is present in flour mills and other places where fine powders can be ignited, and is even greater for fuel tankers and refuelling aeroplanes where

7

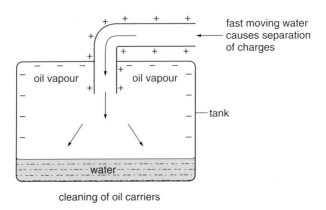

cleaning of oil carriers

Figure 1.2 *Cleaning of oil carriers. When sufficient charge has built up a spark can ignite the oil vapour remaining in the tank.*

fuels generate a lot of charge as they move quickly through pipes. An example is provided by the cleaning process of oil tankers, see Figure 1.2.

An everyday use is illustrated by the ink jet printer, Figure 1.3.

Electrostatic precipitators are charged plates, or wires, in chimneys. As polluting particles leave an industrial process they pass between the plates, which may be charged to tens of thousands of volts. Some particles may already be charged by friction, others will be ionised in the strong field. They are then attracted to the plates where they form a deposit which is shaken or scraped off from time to time.

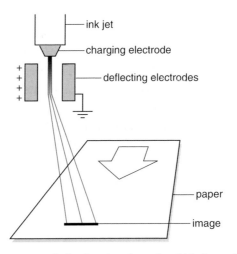

Figure 1.3 *A computer controls the charging electrode which determines where the droplets of ink will be deposited on the paper.*

Spraying items where an even coating is desirable is a third use for electric charges. An example is that of paint spraying. The paint emerges from the sprayer charged by friction. The item to be sprayed can be oppositely charged, or simply earthed, and the paint will be attracted to the item, even to the sides not 'seen' by the sprayer.

- *The quantitative relationship between steady current, charge and time.* Pupils should learn a new unit here, the coulomb (C), named in honour of Charles Coulomb's discoveries, which is a measure of quantity of charge. If coulombmeters are available it is easier to become familiar with this measure and quickly find out that it is possible to build up quite a substantial charge on the meter. Unfortunately the coulomb is a very large unit so the meters are marked in nanocoulombs (nC, that is 1/1 000 000 000 C or 10^{-9} C) but I would not draw attention to this, just satisfy the curiosity of particularly interested pupils who ask. This is reasonable, as pupils would not be expected to manipulate such tricky numbers. Pupils should now be ready to think of electrons as having a fixed charge, though it is a very tiny number indeed, (for teacher information only 1.6×10^{-19} coulombs).

- *About electric current as the flow of free electrons in metals or of the ions during electrolysis.* A demonstration, described in the Activities, involving the electric charge generator will provide a link between static and moving charge, leading to ideas of electric current as flow of electrons. Students should be encouraged to think of a larger current as more electrons flowing in a certain time. The lighting of the neon bulb described in the activities also illustrates that flowing charge is electric current. Having linked electric current with charge flowing, pupils should be ready to accept the idea that the more charge flowing in a certain time, the greater the current. Factors affecting which materials are good electrical conductors can be related to ease of movement of electrons (metals and graphite).

If pupils have met electrolysis in Chemistry (see the Chemistry handbook in this series) this will be a useful moment to review their ideas, if not then a simple demonstration (see Activities) will illustrate the idea that ions can constitute an electric current once they are free to move. Across an electrolytic cell positive ions will move to the negative electrode (cathode) and negative ions to the positive electrode (anode). This may raise the question of which way electrons flow. Note that electrons flow from the **negative pole** to the **positive pole** of a source around a circuit. That constitutes a **positive current** flowing in the regular way.

9

KEY STAGE 4 ACTIVITIES

Electrons

Pupils need to have a feel for an orbital model of the atom, with a massive positive nucleus surrounded by very lightweight negatively charged particles. A football and some little polystyrene balls with charges painted on, showing what is lost and what remains when electrons move about, are useful here. Scale is a problem but several pupils can be enlisted to carry an electron round at some distance and the idea that some are more easily lost, less attracted, becomes clearer. Thus negative charges move, positive charges do not, in solids. Follow up with diagrams of a few different atoms so that pupils draw the correct number of electrons.

Everyday situations

Students will probably remember what electric shocks from the electrostatic machine feel like. Air has a breakdown potential, that is a voltage at which the air becomes ionised and we see a spark. Provide a value for the breakdown potential (say voltage if the word potential is not known) of air, 30 000 volts per cm. Pupils can calculate how many volts difference the machine is creating if they estimate the length of the spark. Advise pupils that high voltages *may* not be dangerous as long as the amount of charge is low, as in the Van der Graaff machine. However they should be advised that mains voltages, though low, transfer a lot of charge and can cause severe burning or death.

Lightning also transfers a lot of charge (separated by friction between moving air and water vapour) so they should know safety rules in this situation too. Do not stand by a tree or in an exposed place. The safest place when you are out is in the car, but do not touch the metal sides as all the charges repel each other to the outside just like the dome of the generator.

Relating current and charge

Students can experiment with coulombmeters by rubbing various rods of materials and touching the plate of the meter. They will soon find out that they can positively and negatively charge objects and can add more and more charge to the meter – it can become quite competitive. I would not record results to avoid the nC problem mentioned in Concepts. To provide the vital link to current electricity demonstrate the following experiment (Figure 1.4) using the electrostatic charge generator. You will need two metal plates on insulating handles and a ball with a conducting surface hanging on an insulating thread.

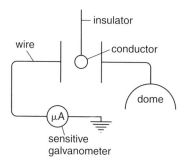

Figure 1.4

You will also need a fairly sensitive ammeter, preferably a demonstration model that can read in milliamps.

The metallised ball swings quickly to and fro between the two plates effectively spooning charge from one to the other and, with careful questioning, this enables pupils to form ideas of electrons flowing along. Since the flow rate (charge per second) can be measured using an ammeter this constitutes current.

How can we make the current bigger or smaller?

Altering the gap between the plates can alter the current, which changes the spooning rate. If it does not work check:

- that the machine is producing sparks
- the plates are not too far apart
- the leads are all functioning
- the ball is on non-conducting thread and hung as far away as possible from the rod of the retort stand

---

If charge per second is current, we can write:

$$\textbf{charge per second} = \textbf{current}$$

$$\frac{\text{charge in coulombs}}{\text{time in seconds}} = \text{current in amperes}$$

This can be written in symbols for more able pupils:

$$\frac{Q}{t} = I$$

A triangle is a popular way of memorising or manipulating formulas. Using it most students will be able to solve very simple problems to find Q or I (see Chapter 5 for the triangle method). Remind pupils that time must be measured in seconds.

Current carried by ions

Remind pupils that the electric currents in wires are carried by electrons. However if positive ions, left when electrons are lost, are free to move they too can flow to create a current. A demonstration of the electrolysis of copper sulphate using carbon electrodes is effective in showing this (with a 6 V direct current supply). Positive copper ions move to the anode where they receive electrons and are deposited as copper atoms.

Granada TV *Physics in Action – Electrostatics 1 & 2* will be a useful resource to use throughout this section.

Assessing pupils' learning

- Provide drawings showing situations with suspended charged balls, rods, balloons and ask pupils what will happen and why. It will help to prevent explanations being given in terms of magnetic poles instead of electrical charges.
- Students could be given calculations using the relationship between current and charge with numbers appropriate to their ability. For less able pupils simple examples using numbers under 10 are appropriate. These are easily written, for instance, if the current in a stage lamp is 2 A how much charge passes in 10 s?
- Dangers and safety precautions can be the subject of research and posters, providing an opportunity for some enjoyably 'shocking' drawings and cartoons. Students could write a leaflet for golfers or kite fliers providing information and suggesting safety measures.

ENRICHMENT AND EXTENSION ACTIVITIES

Some pupils can tackle problems involving currents that are not whole numbers. They should be given a feel for magnitude of current. They can judge whether their answers are sensible or not, for instance 10 A or C/s is a suitable current for an electric kettle whereas 0.25 A would be reasonable in a table lamp.

To encourage an understanding of the link between electricity and magnetism, pupils can observe the effects of an electric field between two charged bodies and compare it with a magnetic field. The investigation can be carried out on an OHP for easy (and more impressive) viewing. Many GCSE or A level texts will describe the arrangement, one suitable reference is given in Resources.

Find out about Benjamin Franklin (one of science's more flamboyant characters who specialised in giving parties of people electric shocks for fun).

RESOURCES

Videos

- Granada TV *Physics in Action – Electrostatics 1 & 2.*
- BBC series Short Circuit called *Shock Tactics.*
- Scientific Eye *Static Electricity* from Yorkshire Television.

Teacher resources

- For an interesting variation on the shuttling ping pong ball see School Science Review September 1997, Vol 79, Number 286, Page 127 *Electrostatic Ping Pong.*
- A good section on electric shock is to be found in *Physics* (Page 278) by Patrick Fullick, published by Heinemann.
- A demonstration of electric fields can be found in the Nuffield Advanced Science Course, Experiment E3. You may need to consult on a substitute for trichloroethane.

chapter 2

Electric current in circuits

BACKGROUND

Electricity and its effects are of course familiar to students. In primary schools they have lots of opportunity to investigate circuits of the bulb and battery type, and SAT tests reveal their increased confidence and knowledge. However they have hazy and often erroneous ideas about concepts like current and voltage, which of these can be shared and what can be said to go around a circuit.

Our model of electric current suggests that:

- electrical charges can move in some materials
- moving charges are what we call current (they can be likened to a flow of liquid)
- as with a flowing liquid, the more charge that flows in a certain time the greater the current
- the charges which move in solids are electrons
- electrons flow more easily in metals and graphite so these materials are called (electrical) conductors
- currents can divide or join at junctions in circuits
- currents flow because a difference of potential energy exists at two points on a conductor
- voltage is a measure of the energy which an electrical source can give to the moving charge. Thus a 1.5 V battery provides less energy to each charge than a 9 V battery

Students will commonly believe that the current 'starts out' from the source taking time to reach the further components and getting 'used up' on the way. The former can be tackled by modelling the current in a circuit as a closed loop of water with a pump in it. When the pump is switched on the water flows everywhere at once, nowhere do you have to wait for the water to flow. In the same way, you don't have to wait for electricity to arrive from the power station when you switch on.

The second idea is probably best dealt with practically (see Activities). Current does not diminish around a circuit. What does diminish is the amount of energy which the charge possesses. As charge passes through each component in a circuit some energy is lost. This loss is attributed to the collisions with the metal ions experienced by the moving electrons. When charge arrives back at the electrical energy source it has no energy relative to the negative electrode, so replaces its energy from the source. There are lots of analogies for this, most of which breakdown if you try to extend them too far. Two are suggested in the Activities, or consult your school's textbooks. Be selective according to the potential of your group and stress that they are models we use because we find it hard to describe electricity.

Ohm's law is a relationship between the current and voltage for a specific component. In practice it is true only for conductors at constant temperature. At Key Stage 3 it is enough for pupils to realise that bigger voltages give bigger currents as long as the metal does not get hot. You can add that Ohm was a school teacher who made this discovery in 1827.

Electricity is very safe if we understand that metals, carbon, solutions (including rainwater and damp skin) are all capable of passing large currents. This must be understood when following safety procedures. We must show pupils that we know about and use appropriate devices, such as circuit breakers, insulated appliances etc. Anything exceeding 35 volts must be regarded as a hazard and no unprotected connectors used. Do not allow the connection of two power packs for any purpose and, if your school still has a supply offering D.C. at voltages below 1000 volts, do not use it as the amount of current available will be harmful.

Electricity is a major piece of work in the curriculum, but not usually a favourite. It would be wise to divide the work into digestible pieces of 2–3 weeks interspersed with non-electrical topics. Concepts can be absorbed before being reviewed and then extended. This way pupils do not look back on, say, a term of work they have perceived as hard and develop an 'electricity phobia'.

Common misunderstandings
- current gets weaker as it moves around a circuit
- voltage goes through a conductor
- potential difference can not exist unless a current is flowing

Pupils should be taught:

- *How to measure current in series and parallel circuits.* Students will already be familiar with the representation of circuits by the use of symbols. It is important that they use up-to-date symbols correctly and can translate them into working circuits. A list of the symbols you will need to begin with are given in most standard physics textbooks.

   We say that the current flows from the positive terminal to the negative terminal around the circuit. The instrument for measuring current, the ammeter, is a very low resistance instrument. When placed in series in a circuit it does not reduce the current flowing. Digital meters are most easily read by pupils but, if they are not available, have some large clear diagrams to pin up explaining how to read difficult scales. If shunts are required make sure pupils understand that a shunt must *always* be used putting in the least sensitive first as a precaution. The unit for current is the amp (A). In full this is written ampere, named after a French scientist.

   In a series circuit the current is the same at all points. In a parallel circuit the current divides at junctions but reunites before reaching the power supply, thus the current leaving the supply is the same as the current returning to it.

- *That the current in a series circuit depends on the number of cells and the number and nature of other components.* In general when more cells are added to a specified circuit the current increases. The size of the current is determined by the resistance of the circuit. If this stays constant then doubling the number of cells doubles the current and so on. However note that:
   a) If any component gets hot its resistance will change.
   b) Some components have a resistance which changes due to other factors such as light.
   c) The resistance of some components depends on the direction of the current.

- *That current is not 'used up' by components in circuits.* This is a difficult concept to dislodge and is best illustrated through investigation. Ammeters show that the current returning to the source is the same as that leaving, whatever components are included. What has been reduced is the potential energy of the charge. As an analogy, as water flows downhill it loses potential energy but the amount of water does not diminish. The quantity of water is analogous to the quantity of charge.

- *That current is flow of charge.* This concept can only be dealt with after work on electrostatic charge. Students seem happy to accept that the charges they generate by friction can move in some circumstances. Those who have felt a shock when touching a charged Van der Graaff machine will certainly describe what they felt as something going through them, something which can even be passed through to other pupils. Good electrical conductors are materials that allow charge to pass easily through them, insulators do not.

Note that the charge flowing is **negative** and so is repelled from the cathode of the battery around the circuit to the anode. This means that charge flows in the opposite direction to that accepted as 'normal' current. This results from an arbitrary decision to call current positive (attributed to Benjamin Franklin) but need not cause any difference in the way we deal with circuit problems.

KEY STAGE 3 ACTIVITIES

Starting circuits

Ask pupils to light a bulb given one cell and one wire, and then to draw a diagram of their arrangement. This will reveal their level of knowledge and memories of Key Stage 2, providing you with a starting point. You can then move onto more practice in setting up circuits or in drawing them as necessary. A CD-ROM called *Crocodile Clips* provides some useful activities here, particularly for anyone with poor drawing skills.

Drawing circuits

Technically a battery is a group of cells though in practice the distinction need not be made up to Grade C at GCSE. In diagrams circuits are generally constructed as rectangles. The long vertical line is the positive terminal of the cell. For consistency draw the cell at the top of each diagram with its positive terminal to the left. Use a red coloured lead from the positive terminal of the cell and the black coloured lead to the negative terminal. This helps in connecting the ammeter correctly (side marked + to positive) or any other devices for which polarity matters. Junctions are often indicated by a dot.

Setting up circuits

- Demonstration of any 'new' components helps to give confidence particularly where expensive items like ammeters are being used. Always have a circuit drawn ready. Ensure students get used to drawing circuits before setting them up or, to avoid boredom, have some ready drawn from which pupils select to stick in their books.
- **S** • Use appropriate safety procedures when doing demonstrations. Always use a

push switch with batteries to avoid them draining quickly. Always switch off a power pack while adjusting a circuit in order to avoid surprises. After all, they do run on the mains supply and, for this reason, can be regarded as more hazardous than a battery. When using the D.C. terminals connect with the correct colour lead even though you can not continue this all round the circuit.

- Circuit boards are still in use in some schools. I would avoid them if individual components are available.
- Explain how to hold the connectors by the terminal not the wire, and not to return wire or items with long leads in a tangle as broken parts lead to a lot of frustration. Hopefully your school will have a system for easily checking in components, especially crocodile clips and magnets!
- If you use power packs for this work, then explain to pupils that they should use the red and black terminals and 3 V for most work.
- Last but not least, have a routine for students to sort out non-functioning circuits. It is vital for your sanity and good practice in any case. This can be given on an instruction card to be borrowed if a problem arises.

Routine for non-functioning circuits

Is the circuit fully connected with no gaps?
Are the bulbs firmly in their holders?
Check your bulbs by borrowing a working one and trying it in your circuit.
Check the batteries with a battery check board (home-made will do).
Check connecting wires by replacing in turn.
Finally ask for help.

Suggested circuits

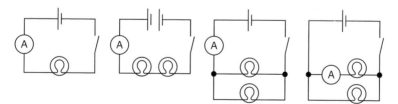

Figure 2.1 *Measuring currents in series and parallel circuits.*

Demonstrate circuit A (Figure 2.2). Add cells one at a time up to 8 (a maximum of 12 V) but only switch on for a short period of time to see the effect on the bulb and the ammeter. Pupils can use circuit B, adding one lamp at a time and recording the current.

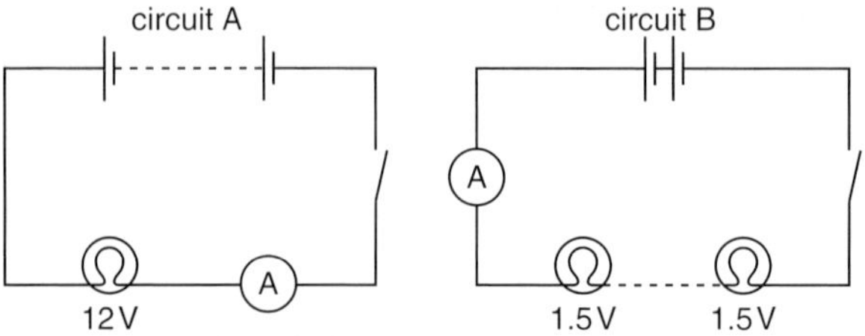

circuit A

circuit B

A

A

12V

1.5V   1.5V

Figure 2.2 *Factors which determine the current.*

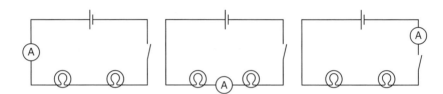

Figure 2.3 *Does the current get used up?*

Pupils should predict the current (higher! lower!) when the ammeter is placed in different parts of the circuit (as shown in Figure 2.3).

Students enjoy investigating other components such as heaters, buzzers, motors, LEDs, LDRs etc. and can appreciate some of their properties without a need to understand the way they work. Provided you limit the number of cells to two, or the number advised by the manufacturer of the kit components, no damage should arise. It would be wise to protect motors by making sure pupils do not stop them turning while a current is passing through.

Flowing charges can be demonstrated through an animated cartoon. A *Science in Action* program called *Electricity* is a good example; information is available on the BBC website (www.bbc.co.uk.education).

Assessing pupils' learning

- Design a *Safety with Electricity* poster/booklet after seeing a video, or a class brainstorm of ideas.
- Investigate the effect on current of using different lengths of resistance wire in a circuit – opportunity for a fair test, quantitative work, graphs. Power packs would be needed here rather than batteries and a warning about wires getting hot.
- Word search of electrical terms.

- More able students or those who have covered electricity in detail before, should extend their ideas about current by measuring values in parallel circuits and discovering that they can be added to give the current which leaves/returns to the battery.

ENRICHMENT AND EXTENSION ACTIVITIES

Find out about animals that can 'make' electricity e.g. electric eels.

Make electricity from fruit etc.

Provide some colour filters, sweet wrappers will do if necessary, and ask students to set tasks for each other such as 'Design a circuit to light either the red bulb or the blue and green bulbs.' Enthusiasts might design a theatre set.

They should also be introduced to the concept of resistance as *the larger the resistance the smaller the current,* and that the resistance of a series circuit is the sum of all resistors in the circuit.

Investigate a rheostat.

Students not ready for this level of difficulty yet should be given lots more practice in drawing and naming symbols and circuit diagrams. These should be given some relevant contexts e.g. a buzzer for a baby alarm, lights for a disco, heater and light for a pet snake habitat etc.

21

KEY STAGE 4 CONCEPTS

Pupils should be taught:

- *That energy is transferred from batteries and other sources to other components in electrical circuits.* The energy per unit of charge available from a battery is the value written on it in volts (V). The energy is transferred to components in the circuit and ultimately is dispersed to the atmosphere. The higher the resistance of the component the more energy is transferred for a given current. Continuing the water analogy, a river's height above sea level decreases much more rapidly when flowing down a steep hill (high resistance component) than when flowing down a shallow one. The pump (battery) renews the potential energy of the water (charge) by raising it uphill (raising its potential difference).
- *That resistors are heated when charge flows through them.*
- *That electrical heating is used in a variety of ways in domestic contexts.* The energy transferred to a component in a circuit depends on both the current and the resistance of the component. Some items such as irons, electric grills etc. operate with a high current so that a lot of energy is transferred in the

form of heat energy. All electrical devices transfer heat energy but in low current devices it is small. For sensitive items excess heat may have to be removed by various methods, such as fans in computer drives.

- *The qualitative effect of changing resistance on the current in a circuit.* For a given potential difference between the ends of a component, the greater its resistance the smaller the current. When pupils describe this make sure they avoid *stronger* or *faster* currents.

- *How to make simple measurements of voltage.* The voltmeter is used first to check the voltage of batteries. This helps pupils to realise that a potential difference can exist when there is no circuit. Introduce it as a diagnostic tool, so that it never appears to be part of the circuit but something that can be clipped on after completion of a circuit to measure the *difference* in voltage (potential) between two points on the circuit. These can be either side of a battery in a circuit to measure the energy available or between *any* two other points, even two points on a piece of connecting wire. The highest differences in the circuit will be where the most energy is being transferred. The voltmeter is a very high resistance instrument so if it is incorrectly placed in series pupils will immediately conclude 'the circuit's not working'. If you have meters requiring shunts choose a shunt which is greater than or equal to your supply.

- *The quantitative relationship between resistance, voltage and current.*

- *How current varies with voltage in a range of devices, including resistors, filament bulbs, diodes, light-dependent resistors (LDRs) and thermistors.* The law, known as Ohm's law, tells us that the current through a fixed resistor increases in proportion to the voltage applied. Written in mathematical form,

$$\frac{\text{voltage}}{\text{current}} = \text{a constant}$$

The constant shows the resistance of the resistor to allowing current to pass and we can say $\dfrac{\text{voltage}}{\text{current}} = \text{resistance}$, for voltage in volts, current in amps and resistance in ohms (Ω). The triangle method may be used (see Chapter 5). Pupils will have to accept that 'I' is the symbol for current. Note that this law only applies to a fixed resistor. The fact that a current is passing causes a temperature rise which, in metals, usually increases resistance. So-called resistance wire has a reasonably constant resistance over a useful current range, but anything getting hot like a heating element or light bulb, will show a curve for a graph of I/V. This is shown in Figure 2.4 together with the characteristic curves for other components.

- *That voltage is the energy transferred per unit charge.* Voltage as potential difference is a difficult idea for students. The latter term is only appropriately

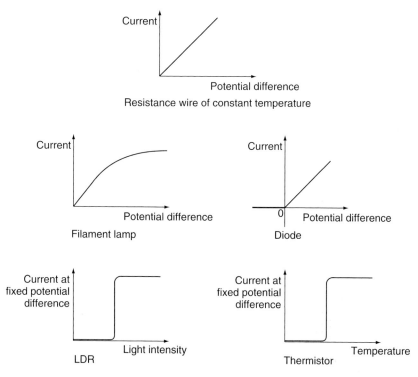

Figure 2.4 *The LDR and thermistor behave like switches.*

used with able students in Years 10 and 11. For the majority the term voltage is sufficient. Voltage or potential difference should be thought of in terms of energy, energy which is given to charge by a source of electricity and which reduces as current passes through components.

$$\text{voltage} = \frac{\text{energy}}{\text{charge}} \text{ J/C}$$

- *The quantitative relationship between power, voltage and current.*
- *How measurements of energy transferred are used to calculate the costs of using common domestic appliances.* To make sense of this relationship students should think about voltage as **energy per unit of charge** and current as the **flow of charge**. Increase in either of these factors will increase the power being taken from the supply. The product of the two is the power measured in watts (W).

power (in watts, W) = voltage (or potential difference, in V) × current (in A)

The triangle may be used; see Chapter 5. Note that 1000 W is 1 kW (1 kilowatt).

If a component with a power rating of 1 kW is used for 1 hour the energy transferred is said to be 1 kWh (1 kilowatt hour).

Example: What is the cost of running a 2 kW fire for 3 hours? Assume 7 pence per unit.

$$2 \, kW \times 3 \, h = 6 \, kWh,$$
$$6 \, kWh \times 7 \, p/kWh = 42 \, pence$$

- *The difference between direct current (D.C.) and alternating current (A.C.).* Direct current is one where the current flows consistently in one direction. You would get this from a battery or power pack using the red (+) and black (−) terminals. Some components are not 'fussy' about which way the current flows but some are, such as buzzers and diodes. Alternating current flows to and fro. If you are generating it with a dynamo the frequency with which it changes direction depends on how fast you turn the handle/pedal. If it is taken from the mains, such as a power pack, the yellow terminals, the frequency is fixed at 50 cycles per second. The symbol for an A.C. supply is ~.
- *The functions of the live, neutral and earth wires in the domestic mains supply, and the use of insulation, earthing, fuses and circuit breakers to protect users of electrical equipment.*

1 The live wire, covered in brown, is connected directly to the power station when all the switches in its circuit are on. For this reason any safety device must work to disconnect this line.

2 The neutral wire, blue, carries current back to the power station but is maintained at Earth potential so that charge does not need to 'find' a path to Earth.

3 The Earth wire, yellow and green, is connected through some path to Earth so is not a danger. It provides a path to Earth for excess current which may arise from a failure such as a short circuit or a bare live wire.

4 The wire covering is plastic, a very good insulator, and a cable has an outer layer of plastic. This must be held in place in a plug by a cable grip.

5 A fuse is a small piece of thin wire, in a cylinder, possibly containing sand. If more than a chosen current flows, the wire melts and the circuit is broken. It must be set in the live wire so that it will cut off the current if there is an overload.

6 Simple circuit breakers use an electromagnet to attract a piece of flexible steel when there is an overload. As the steel is part of the circuit its

movement will cut off the supply. They are more responsive, quicker and more convenient than fuses as they can be reset.

7 Appliances are usually insulated with a double layer of plastic as well as having Earth connections.

KEY STAGE 4 ACTIVITIES

Resistance

After reviewing work from KS3, students should become familiar with the idea of resistance by setting up a series circuit with a power pack, an ammeter, a fixed resistor, say 5 Ω, to protect the ammeter and resistance wires, normally alloys such as constantin or nichrome. They should investigate the effect on the current of changing the length, thickness and material of the resistance wire. Discuss the meaning of resistance. Describe it as the ratio of voltage applied to current produced and give the units. Challenge students to produce two pieces of wire of different materials, thicknesses and lengths that have the same resistance. Allow them to investigate rheostats and 'pots' (potentiometers), asking for an explanation. They are used in dimmer switches etc. See Figure 2.5.

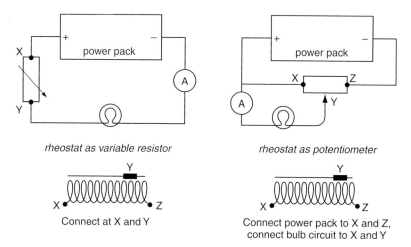

Figure 2.5 *Circuits showing a variable resistor and a potentiometer.*

S Students will notice that short pieces of wire become hot (give a warning beforehand) and should relate this to bigger currents. Plenty of useful examples can be given for this heating effect but don't forget that it can be wasteful and a nuisance too.

Voltage

The voltmeter should be demonstrated to students by first building a series circuit containing bulbs or other components which can be seen to be working. The voltmeter can then be attached to two points in the circuit and will record the difference in voltage between them. Students can then be given such tasks as:

1 Check the voltage across 1 cell, 2 cells, 3 cells, 3 cells where one faces the 'wrong way' (predict first).

2 Check the voltage across the power pack set on D.C. – does it match its reading?

3 Check the individual voltages across two lamps in a) series, b) parallel – what do you notice?

(In series they should add to give the voltage supplied to them, in parallel they should read the same as each other and as the supply unless there are other components.)

Ohm's law

A demonstration of the verification of Ohm's law should be done with students helping out with the readings and plotting a graph, preferably on an OHP. If current (vertical) is plotted against voltage (horizontal) the relationship can be shown as *the greater the energy supplied to the charge the larger the current.* However if you want to calculate resistance you will need to plot the axes the other way round or you will end up needing to find the reciprocal of the gradient i.e. voltage/current (which is much too confusing). Using the same results but plotting the other way to find resistance could be a homework task; make sure you know the answer for the resistance.

Non-ohmic devices

Students can now be given a range of devices to investigate. These can include a 12 V bulb, LDR, thermistor, reed switch and diode. Discuss their operation and get pupils to suggest where they might be used.

Examples: *Diode*   *in the conversion of A.C. to D.C.*
                Thermistor   *allows a large current when the temperature is high*
                LDR   *allows a large current when in bright light*
                Reed switch   *it can operate a large current when given a 'magnetic' signal from a low current circuit*

Energy and power

Students should able to see rating labels on electrical devices and use the relationship for power to calculate the current flowing. They will find that devices involving heating require much higher currents and, if they then calculate kWh, will be able to see that they are the most costly to run. Demonstrating a domestic meter is useful. Students can see the wheel inside turn faster for some items. Begin with low energy consumers and work towards kettles and toasters. Find out which appliances really use the most electricity in the house. Some schools will have joulemeters which can 'count' the joules of energy supplied to a circuit (see Chapter 14).

Demonstrating D.C. and A.C.

Method 1. Use a demonstration centre-zero meter so that everyone can see. Connect a small (5 Ω) resistance in series with a single cell and the ammeter. A fixed reading will show. Reverse the cell, same reading, opposite side. Then connect an alternator (bicycle dynamo) and wind slowly. The pointer will move correspondingly from side to side showing alternating current. Don't connect up the mains supply to this meter – apart from safety considerations the vibrations of the pointer at 50 cycles/second would probably not cause any movement at all.

Method 2. You will need to become familiar with the Cathode Ray Oscilloscope (CRO) for several experiments so get help if possible from a technician or other expert. Failing that find the simplest CRO model in the department and, before the lesson of course, switch on and adjust the dials until you obtain a dot (Figure 2.6). You can not do any damage by simply altering the settings and it will help you to get to know them. Use the time base dial to get a smooth line. Look for the volts/division knob and turn it to 1 volt/division. Then connect a cell across the input terminals. The result will be one of the lines in the diagram depending on which way round the cell is. Altering the volts/division will change the position of the line vertically. This is a 'picture' of direct current. Attaching the alternator as above will give a 'picture' of alternating current. Instead of an alternator you can attach a power pack on low voltage A.C. to the CRO. This will give a 50 cycle/sec frequency, though not a smooth sine wave. Alternatively you can use a signal generator and choose any required frequency you like. You may need to adjust the time-base dial to obtain a nice trace (Figure 2.6).

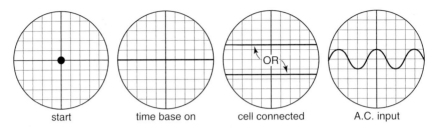

start   time base on   cell connected   A.C. input

Figure 2.6 *CRO trace.*

S Safety

Pupils believe that electric shocks are the greatest cause for concern but they should appreciate that more deaths occur from electrically caused fires than from shocks. The first aim of protection is to prevent contact with live items, wires etc. and in case of an accident to cut off the power supply as quickly as possible. Some demonstrations are possible but a video is probably safer as it is less likely to encourage pupils to try out any of the things you advise against. (See *Powerful Stuff* – CEGB.)

A few strands of steel wool connected across a power pack and resting on a heatproof mat show the principle of the fuse. You may get a small shower of sparks so wear safety goggles and warn pupils that this must not be tried with mains voltage. Note: you can teach that fuses melt or blow but do not allow 'blow up', as it is inaccurate and very unpopular with examiners.

A *Physics in Action* tape has a helpful section on electricity in the home. See Resources.

Assessing pupils' learning

- Lots of practice with Ohm's law for a range of values, from Christmas tree lights to a starter motor for a car.
- Set 'n' solve circuit problems involving Ohm's law, as a group activity.
- Research the way in which an appliance works as appropriate to ability, from torch to fridge, kettle to TV – write a report, maybe as an archaeologist from 2050 AD.
- Calculate and compare running costs of electrical devices – which costs more, Dad making a cup of tea or Kris listening to a CD player for an hour?

ENRICHMENT AND EXTENSION ACTIVITIES

For students who need basic ideas reinforcing:

- cut out pictures of appliances from magazines and identify the conductors and insulators
- draw in wiring on large diagrams of appliances, such as toasters, kettles and plugs
- use sensors with lights, buzzers etc. to build warning lights: has someone opened your bedroom door? Is the incubator warm enough for the baby?

List electrical appliances; compare them with their equivalent, if any, 50 or 100 years ago.

More able pupils should get to know the term potential difference instead of voltage. Difference in electrical potential can be modelled in the same way as difference in gravitational potential (tying in with the second analogy above) but this is only for the very confident.

Study of resistors in series and parallel.

In series the total resistance of a circuit which contains two resistors R_1 and R_2 is simply their sum i.e. $R_{total} = R_1 + R_2$

In parallel the total resistance is calculated from this equation:

$$\frac{1}{R_{total}} = \frac{1}{R_1} + \frac{1}{R_2}$$

If you substitute two values for R_1 and R_2 you will quickly see that the total resistance is less than either of the two. In practical terms, it is 'easier' for the current to get round the circuit when there are two routes than when there is just one.

Compare the cost of using a battery and mains supply to run, for example, a stereo.

Show a demonstration of cathode ray tubes, such as the diode and Maltese cross tube. If you show these, your school will have the proper stand and safe connectors. Make sure you check the manufacturers' instructions so that the correct voltages are applied. Research the link between them and the CRO and television tube.

RESOURCES

Videos
- *Science in Action* BBC (BBC@co.uk.education).
- *Powerful Stuff* CEGB.
- *Physics in Action (Electricity in the Home)*. The tape can be hired from the Institute of Physics at The Education Department, Institute of Physics, 76 Portland Place, London W1N 3DH.
- Scientific Eye *Electricity* from Yorkshire Television.

Information technology
- CD-ROM *Electricity and Magnetism* from Yorkshire International Thompson Multimedia Ltd Tel 0161 627 4469.
- *Voltage and resistance experiments*, Roy Barton, available from Dataharvest, Leighton Buzzard, Bedfordshire.

Teacher resources
- The Open University Professional Development in Education, Unit C Part II.
- *Current understanding* by Summers, Kruger and Mant, published by OUP.

chapter 3

Magnetic forces

BACKGROUND

Magnetic forces can appear mysterious in that they are non-contact, like electrostatic and gravitational forces. The only common element that reacts appreciably to a magnetic force is iron. As this is the main constituent of steel, steel is, therefore, also a **magnetic material**. Nickel and cobalt also have magnetic properties.

Iron-containing rocks such as lodestone have been known since very early times to attract or repel pieces of iron. The Chinese used the ability of some pieces of iron ore to construct early compasses though without attributing the phenomenon to the Earth's influence. It was not until the seventeenth century that the concept of a magnetic field was applied to the Earth.

Magnetic materials can be made into **magnets** with two **poles**, North and South. There is a force on an imaginary isolated magnetic N-pole near a bar magnet. It acts along an imaginary **field line** and is directed towards the S-pole of the magnet. The whole pattern of field lines around and between the poles is called the **magnetic field**. The magnetic force decreases with distance from the magnet, eventually becoming too small to be detected.

If a magnet is suspended its N-pole points in the direction of the Earth's N-pole or, more strictly, the Earth's magnetic North (the magnetic axis is displaced by approximately 11° from the geographic axis). The whole Earth has a magnetic field as though it contained a giant bar magnet buried in the core. Its field lines spread far into space.

A magnet can induce magnetism in a piece of magnetic material and will then attract it. So a magnet placed near a piece of unmagnetised but magnetic material will attract it at both ends.

Magnetic fields also exist around conductors carrying electric currents. This effect is the topic of Chapter 4.

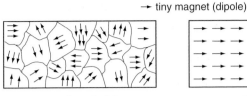

tiny magnet (dipole)

magnetic material
that is unmagnetised

magnet

Figure 3.1

A full and detailed explanation for students of these phenomena is not possible or appropriate but the characteristic properties of magnetic materials can be explained by 'domain theory'. Tiny magnets exist in magnetic materials as a result of unbalanced moving electrical charges. These group together in regions or domains. They are arranged randomly in unmagnetised iron. In a magnetic field they are forced to become aligned. See Figure 3.1. Magnets can lose their magnetism through being hit or dropped as this provides energy for domains to move back to a more random arrangement. The movement of charged particles within the core of the Earth offers a useful explanation of its magnetic field since the core is molten. This is a question which you may well be asked.

Common misunderstandings
- metals are magnetic
- filings is spelt fillings
- confusion between *magnetic* and *magnetised* often arises

KEY STAGE 3 CONCEPTS

Pupils should be taught:

- *About magnetic fields as regions of space where magnetic materials experience forces.* The magnetic field, represented in two dimensions, is shown for a bar magnet in most physics textbooks. In reality it exists in three dimensions, as you would see if you allowed iron filings to stick to a bar magnet (but beware the difficulty of removing them and danger of splinters). Non-magnetic materials, such as paper or one's hand, do not stop magnetic forces. When fields are investigated they are found to be concentrated at the poles of the magnet, that is, they are strongest where the lines of force are greatest in number. Note that the direction of any lines of force should always be marked.
- *About the field pattern produced by a bar magnet.* Free of the presence of other

magnets a compass needle will rest N–S in the Earth's field. Placed near a bar magnet the compass needle will change direction and can be used to plot out individual field lines, building up a complete pattern. The position of the needle is the resultant of a combination of the Earth's field and the magnet's but the Earth's will be relatively weak while you are within a piece of paper's distance of the magnet.

KEY STAGE 3 ACTIVITIES

Magnetic fields

Pupils will be familiar with magnetic toys but will still enjoy playing with them or watching a good one demonstrated. Magnetic toys include ones that show attraction and repulsion, forcing objects to move without being touched (the fishing game), float mysteriously and so on. You can invite pupils to make a simple snake charmer using tissue paper for the snake with a small paper clip or similar at the head end. The magnet can be disguised as a musical pipe and designing/following a template for a snake basket improves skills in volume work.

There are also those which form various patterns, faces etc. with iron filings. The latter should only be available to pupils in sealed containers or used by you on an OHP.

You, or the students, can place some materials (paper, your hand) between attracting magnets to check whether this stops the force. Try various materials including a range of metals. Results may be surprising to students. Quite strong magnets are necessary to be effective and, if students use them, they should be warned to avoid dropping them and to store them with 'keepers' at each end. They are worth counting in, as they will certainly reduce in number otherwise.

A really strong permanent magnet is always fun to see. Invite someone to remove the keeper – this is usually impossible because they try to pull it off directly. You can then remove it easily by sliding it off. You can try out different materials as above and it will also make suspended magnets do interesting things. Do not dip it into iron filings or put it near to a small compass needle or your watch.

While students have a good bar magnet they should try suspending it. Avoid iron/steel clamp stands; use wooden ones or upturned wooden stools. It may be more convenient to float the magnet by placing it in a plastic petri dish on some water in a sink or tray. They should align N–S (check that you know where North is) but, if any do not, look for steel table bars or pipes which might affect them. Ask what problems this might pose for navigation on board ship etc.

Bar magnets

Pupils usually enjoy investigating field patterns, once they have grasped the technique. You can choose some to be mounted and put on display. I would not discourage pupils from using colour as long as there is no shading and no lines *ever* cross.

Firstly they should arrange the compass away from any magnets so that it is pointing N–S. They should bring up a magnet to see at what distance the compass needle is affected. This can be seen as a measure of the strength of the magnet – useful later for investigating electromagnets. To form a more complete picture of the direction of the force at any particular place suggest that the bar magnet is kept still and the compass moved. Some idea of field lines will emerge. The whole magnetic field can then be plotted very carefully. The method of doing this needs explaining thoroughly. If you have a small compass which is transparent top and bottom the best way is to use an OHP. Proceed as follows:

- Place your magnet in the middle of a piece of plain paper. Draw around it in case it gets moved.
- Place a small (plotting) compass near the magnet and mark a dot on the paper at each end of the needle.
- Move the compass so that one end of the needle points to the dot, which marked the other end. Make another dot.
- Continue in this way until you return to the magnet or reach the edge of the page.
- Join this row of dots (field line) with as smooth a curve as possible.
- Repeat from another point on the magnet to give a spread of lines all over the page i.e. the field. Each line should carry an arrow directed towards the S-pole.

It is worth spending plenty of time so that everyone manages to produce a good drawing. Fast workers can be given various arrangements of two magnets or a horseshoe magnet.

As an alternative or backup the whole magnetic field can be shown on an OHP using iron filings, but sprinkle them lightly on a transparency over the magnet so they do not stick to it. Remove with care.

Assessing pupils' learning

- Provide some diagrams which show field lines and ask pupils to mark in the N and S-poles. You can include a trick one, that of an iron ring which has circular field lines. In this case it will not be possible to say where the poles are located and magnetic objects will not be attracted towards it.
- Design an alarm to include a magnet, which will ring a bell if a safe door is opened. Discuss the circuit in groups first.

- Describe a compass you could use to help find your way over a moor. Mention any problems that might cause misleading results.
- Design a separator for recycled steel and aluminium cans.

ENRICHMENT AND EXTENSION ACTIVITIES

Pupils can draw magnetic fields with neutral points, that is, places where the resultant magnetic force is zero. For instance a neutral point is to be found between two like poles facing one another. Try also opposite poles facing, magnets at right angles, three magnets close together etc.

They should be able to distinguish magnetically soft materials (easily magnetised and demagnetised) and hard magnetic materials. The former includes so called soft iron from which electromagnets are made. The latter includes steel, one material from which permanent magnets are made. Allow them to investigate assorted nails, pins, tacks, paper clips and so on to try to achieve a scale of magnetic hardness.

Some pupils will find the concept of field lines difficult and might practise using the idea by being given clear drawings of fields and asked to draw in compasses pointing the correct way.

KEY STAGE 4 CONCEPTS

Pupils should be taught:

- *That like magnetic poles repel and unlike magnetic poles attract.* This concept will have emerged from the earlier work and is not a difficult idea, provided that the words used are explained. It is almost instinctive to feel that attraction is the crucial factor in testing for a magnet. While this will distinguish *magnetic materials*, you can only be sure that you have a *magnet* if you observe repulsion with one pole of a known magnet.
- *That a force is exerted on a current-carrying wire in a magnetic field and the application of this effect to simple electric motors.* It is good to present the link between electricity and magnetism as the amazing phenomenon that it is. All wires carrying a current, have an associated magnetic field whose pattern depends on the shape of the conductor. Three simple shapes are shown in Figure 3.2 with their fields.

If there is a magnetic field then there will be associated forces. Thus, just as there is a force between two magnets, there will be a force between one magnet and a current-carrying wire and also a force between two current-carrying wires.

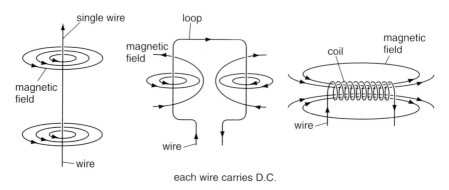

single wire

loop

magnetic
field

magnetic
field

magnetic
field

coil

wire

wire

wire

each wire carries D.C.

Figure 3.2 *Each wire carries D.C.*

The force on a loop of wire in a magnetic field is relatively small but can be increased by:

- increasing the current through the wire
- using more loops or turns of wire
- placing the coil in a stronger magnetic field.

The effect is a strong enough force to turn the coil provided, of course, that it is free to turn. This is the principle of the electric motor. Please note that in Figure 3.2 the arrows on the wires represent the direction in which the current flows.

KEY STAGE 4 ACTIVITIES

Repulsion/attraction/induction

Like poles repel, unlike poles attract. If appropriate use like = alike, repel = push away, unlike = opposite.
    Ask pupils:

- *How can you persuade a piece of steel, such as a nail or paper clip, to become attracted simply by putting a magnet near it?* Use the idea of regions of tiny magnets lining up.
- *Does the nail now act like a magnet? How do you know?* (By its effects on other nails or a compass needle.)
- *How many induced magnets can you persuade into a line?* This is a measure of the strength of the magnet.
- *Does your 'new' or induced magnet stay magnetised when the permanent magnet is removed?* For steel the answer is yes because steel is magnetically hard. For iron the answer is no because iron is magnetically soft.

In practice iron nails tend to retain their magnetism so be aware, particularly if you are trying to show the nails falling when a magnet is removed.

Puzzle objects can be used to distinguish magnetic materials from magnets and to underline the need for *repulsion* as a test for a magnet. Pupils can be given, several identically covered bars. Find out if they are:

- non-magnetic metals
- magnetic materials
- magnets

They can not take off the wrapping or use iron filings to help them. You may not be able to supply convincing identical bars but the exercise can be carried out as a thought experiment.

Forces on wires

To demonstrate the force on a conductor carrying a current in a magnetic field a wire can be suspended so that it is free to swing between the poles of a large magnet. The wire is part of a circuit with a cell and switch, but the contacts need to be just touching to allow the wire to move. When you switch on the wire will 'kick' showing that a force is acting. All bare wires will make better contact if rubbed with sandpaper before connecting up. See Figure 3.3.

A useful kit to demonstrate all the electromagnetism in the curriculum, which is available in most schools, is the Westminster kit. Use the magnets from this kit which are strong and have their poles on the flat faces instead of at the ends. Be aware that the magnets are brittle and should be handled without being banged together. Attach the magnets to the yoke, arranged so that they are attracting one

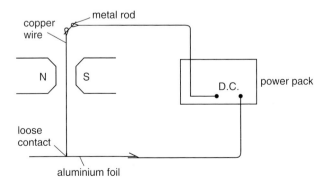

Figure 3.3

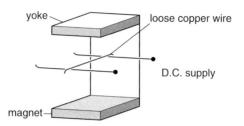

yoke

loose copper wire

D.C. supply

magnet

Figure 3.4

another. Use two parallel copper wires to connect with a third piece, which is loosely laid across them, see Figure 3.4.

When you switch on the power pack the wire jumps, and usually breaks the circuit. This is good because you do not want the large current to flow for any length of time.

- *How can you make it jump the other way?* Investigate reversing terminals, magnets, both. Each will reverse the force so both will restore the original.
- *What is the direction of the force compared with the wire and the magnetic field?* The three are described as mutually at right angles which means the directions follow the lines of wall and floor when they meet at the corner of a room. Show pupils how this can be replicated with thumb and first two fingers. If you want to find the direction of the force you point the First Finger in the direction of the magnetic Field (not forgetting it goes N to S), the seCond finger in the direction of the Current then the thuMb gives the direction of the Motion. For this to give the correct answer it is essential to use the *left hand,* which gives its name to the rule.

To show that there is an opposite force on the magnet, place one in a petri dish and float it. Hold the wire carrying a current over it as in Figure 3.5. The magnet always rushes away. Ask pupils to predict the direction of movement using the left hand rule. Don't forget the force on the wire is equal and opposite to the force on the magnet.

The electric motor

Once movement has been observed, lead students into the idea of utilising the force by arranging the wire, not just to jump off or swing, but to rotate so that the force can be supplied continuously. Clear diagrams of the electric motor can be found in school texts. Understanding is greatly enhanced by actually making a working motor. The Westminster kits are very effective and can be made to spin rapidly. The weakest point is the split ring which is difficult for less nimble fingers to assemble. It helps to have suitably long pieces of wire ready cut for ten

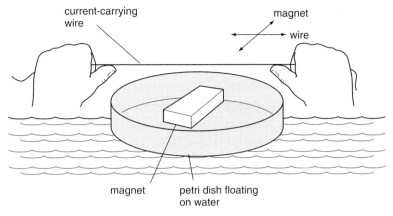

current-carrying wire

magnet

wire

magnet

petri dish floating on water

Figure 3.5

turn coils, with at least 3 cm bare ends. If time allows prepare a working motor in advance and keep available for students to check against. Give pupils who did not get their own motor to work a chance to try out one of the successful ones.

All students will need help in understanding just how the rotation of the coil is caused by the force acting on the two sides of the coil which are perpendicular to the field lines of the magnet. They will also need to be shown that, not only is there zero force when the coil is in a vertical position (see a diagram in any standard text) but to maintain the rotation in the same direction the current must be reversed each half turn. This is achieved by the split ring and brush-like contacts (a device called a commutator).

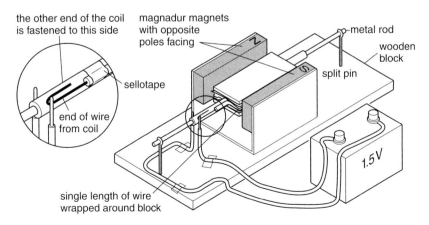

the other end of the coil is fastened to this side

magnadur magnets with opposite poles facing

metal rod

wooden block

sellotape

split pin

end of wire from coil

single length of wire wrapped around block

1.5V

Figure 3.6 *Simple electric motor.*

Students will be able to suggest ways of making this force even bigger but will need guidance to appreciate how it can be put to use as a motor by winding the wire into a coil so that it can rotate. Making a motor can be a trying exercise, particularly the fiddly connections, but I always feel it is justified by the satisfaction achieved by successful groups. If you judge your group would benefit more from demonstration, partially construct a motor before the lesson. Students must be shown the connections to a motor consisting of split rings partly because the wires would get tangled as the coil rotates, but particularly because the direction of the current must change every half cycle to keep the coil rotating in the same direction.

Assessing pupils' learning

Ask pupils to make a list of ways of:

- making a motor turn faster
- making it turn the opposite way

ENRICHMENT AND EXTENSION ACTIVITIES

To increase familiarity in relating the simple motor to the machines they use everyday, pupils should look for examples at home and find out about their power supplies and any dangers associated with their use. Anything from hairdryers to chainsaws can be the subject of research using books or CD-ROM – no practical work though.

Pupils should suggest and research ways of increasing the speed and load that a motor can lift. A demonstration taken from an old suction cleaner or electric drill with the outer casing removed is useful. They will see that there is not just one coil but many at different angles on the axle. The magnet providing the field may also be an electromagnet. It may be adapted to operate on A.C.

If efficiency has been covered, pupils can design and/or carry out an investigation into the efficiency of an electric motor by lifting a small load. They can find out the efficiency of fuel engines for comparison. See Chapter 15 for details of the calculation of efficiency.

Research the working of an ammeter or a loudspeaker, resources are suggested below.

RESOURCES

Video

- Scientific Eye *Electricity and Magnetism* from Yorkshire Television.

Information technology

- CD-ROM *Electricity and Magnetism* from Yorkshire International Thompson Multimedia Ltd Tel 0161 627 4469.

Teacher resources

- For diagrams and photographs showing electric motors, loudspeakers and ammeters (galvanometers) see *GCSE Physics* by Tom Duncan, Sections 58 and 59.
- Written for teenagers with lots of investigations you might use, *Magnet Science* by Glen Vecchione. (ISBN 0 8069 0889 0, published by Sterling Publishing. Importers are Chris Lloyd Sales and Marketing Services, 463 Ashley Road, Parkstone, Poole BH14 0AX.)

chapter 4

Electromagnets

BACKGROUND

At Key Stage 3 pupils are generally unaware of links between magnetism and electricity, even if they have had the opportunity to make electromagnets. This is a chance to show new phenomena and bring out many important applications from microphones to motors and from transformers to generators. A very important figure in the early development of electromagnetism is Michael Faraday and the experiments suggested below are relatively unchanged since his demonstrations at the Royal Institute in the 1830s.

The lack of familiarity and introduction of a highly technical vocabulary combine to make this topic the least well understood in the whole GCSE curriculum. The following represents an order for presenting the material, so that pupils can build up their knowledge layer by layer. It is not intended to be taught as one continuous section.

1 The magnetic fields of two magnets interact showing repulsion between similar poles. See Chapter 3.

2 The magnetic fields of two wires carrying a current interact, showing attraction between currents flowing the same way.

3 Whenever an electric current flows (equivalent to a charge moving) a magnetic field is associated with it. The pattern of magnetic field lines is dependent on the shape of the conductor, such as a straight wire, a long coil etc., see Figure 3.2. This field has the same characteristics as the field of a permanent magnet but has the added advantage that it can be switched on and off. The strength of the field depends on the current flowing, so inconveniently large currents would be needed to produce substantial field strengths. However if the field is used to induce magnetism in a piece of soft iron the field is intensified. This, too, can be switched on and off, at high frequency if required. The device is an electromagnet.

4 The magnetic field produced around a coil can interact with the field of a permanent magnet obeying the rules of attraction and repulsion. As we can not easily tell where the North and South poles of the magnetic field of a conductor are, there are rules to remind us which way the magnetic field lines go in relation to the direction of current. The electric motor was detailed in the previous chapter and involved an input of electrical energy and achieved an output of kinetic energy. Energy considerations suggest that it should be possible to input kinetic energy and achieve electrical energy. The device in which this is done is called a dynamo or, on a larger scale, a generator. This is shown when the coil of a motor, as constructed in Chapter 3, is turned. A potential difference is created between its terminals. No potential difference arises if the coil is stationary.

Example: When turning the coil by pedalling, a voltage is produced, enabling a lamp to be lit. If you stop pedalling, the light goes out.

5 The reason for the induced potential difference (or electromotive force) is the interaction of the moving conductor and the field lines of the permanent magnet. It happens whether the conductor is moved, the magnet is moved or the current switched on or off. However, the permanent magnet could equally well be replaced by a coil carrying a direct current, since this has its own magnetic field. This could be applied to the motor or the dynamo.

6 In stage five we have two coils, one carrying a direct current to create a magnetic field and one turning to cut through the field lines. If an alternating current is used in the first coil then the field experienced by the second coil is changing without the need to turn it. This will induce a potential difference across the second coil. It is possible to apply one potential difference to the first coil and obtain a potential difference in a second coil that is much higher or much lower. The device based on this principle is called the transformer. You will see that this crucial discovery enables the efficient distribution of electricity through the National Grid.

Common misunderstandings

The whole difficulty with grasping the links between magnetism and electricity is rather greater than the sum of the difficulties encountered in either topic alone. A wall chart of terms used with explanatory diagrams is a useful item for reference.

KEY STAGE 3 CONCEPTS

Pupils should be taught:

- *That a current in a coil produces a magnetic field pattern.* The field patterns are shown in Figure 3.2. When formed into a long coil the field resembles that of a bar magnet with the addition of a strong uniform field inside the coil.
- *How electromagnets are constructed and used in devices e.g. electric bells, relays.* An electromagnet is a coil (solenoid) with a core made of a magnetically soft material such as soft iron. There is only a magnetic field when the current flows. The electromagnet can respond very quickly to being turned on and off, as in the electric bell. See Figure 4.1.

KEY STAGE 3 ACTIVITIES

Fields and currents

You can arrange a suitable coil for demonstration on the OHP by winding a coil on something transparent like a test-tube. Place it on the wrong end of a clamp

45

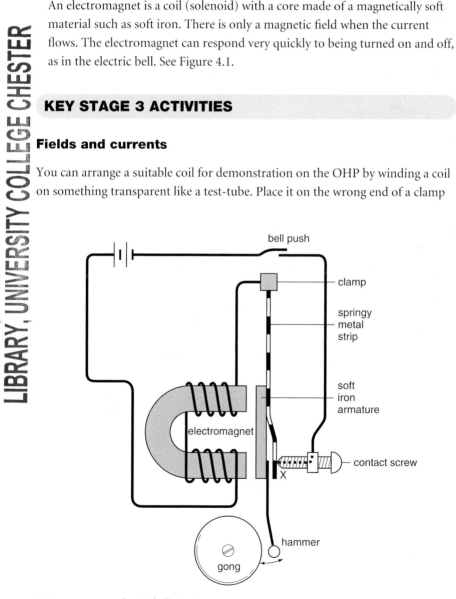

Figure 4.1 *An electric bell circuit.*

so it is suspended over a transparency. A low-volt/high current power pack is a suitable power supply. Sprinkle iron filings over and round the coil. Some will stick to the coil, show them falling off when you switch off. Introduce a soft iron core and note the increased effect. This may not lose its magnetism when switched off as you might have expected. The demonstration is worth repeating with plotting compasses (asking for predictions first) as they will show direction as well.

- *How could we reverse the direction of the field lines?* Reverse the current or turn the tube round.
- *How could we increase the effect?* Increase the current, wind on more turns, move the tube nearer.
- *How could we measure the strength of the magnet we have made?* We can not get an absolute measure of field strength at this level but we can investigate the variable comparatively.

Investigating electromagnets

Finding out how a particular variable affects the strength of an electromagnet is a good subject for investigation allowing for differentiated outcome. Pupils should have a range of equipment available to them so that they can select their own variables from current, turns, cross-sectional area of coil and any others which are safe to do. Again use the low volt/high current packs with a variable resistor. Restrict wire length to a *minimum* of 20 cm and remind pupils to switch on only for a short time. It is best for investigators to work towards stronger magnets rather than weaker as magnetism is often retained, so gives confusing results. They can judge the strength of the field in several ways:

- counting the number of small pins/paper clips picked up
- measuring the length of a chain of hanging pins
- measuring the angle of a plotting compass needle at a fixed distance
- measuring the distance at which the field can just be detected by a plotting compass
- measuring the force with a low reading force meter – this needs to have the magnet in a vertical position
- measuring the repulsive force on the plate of a top pan balance, also with a vertically placed magnet.

The latter two are probably best left for pupils who could do this investigation at Key Stage 4.

Using electromagnets

Give pupils a reed switch and a magnet (weak ones will be satisfactory here) so that they can see and hear them working. They make an audible click. The magnet should be placed alongside the reed, not end-on which would cause repulsion between the contacts. If you can get hold of a large relay from a mechanic it is much easier to see and hear.

The electric bell circuit is given in Figure 4.1 but it is worth having a diagram which looks like the one you are demonstrating. The layout can vary. Students should be aware that its operation is a cycle, a make-and-break circuit. If during the demonstration nothing appears to happen then change the batteries as they are often left in place for a long time and may deteriorate. If there is a small spark or an obvious attraction of the armature but no ringing try adjusting the contact screw or the gong. Doing this in the lesson can be useful if pupils are asked for ideas and their suggestions considered and tried.

Assessing pupils' learning

- The investigation on the strength of electromagnets forms a useful assessment for Sc. 1 in this section.
- Ask pupils to write about the use of magnets in a scrap heap. Which objects from a scrap heap (named items) could be picked up by a magnet? What type (permanent or electromagnet) would a scrap dealer prefer to use and why? You may get arguments about cost here as well as convenience – the class can discuss.
- Draw a flow chart explaining the operation of the electric bell.

ENRICHMENT AND EXTENSION ACTIVITIES

Pupils can reinforce knowledge by answering simple questions involving diagrams of electromagnets. They can identify by colour the parts where current flows, where field lines are, what another magnet would do if it were brought near etc. This can also be done with diagrams of the bell and relay.

Design a level crossing gate, using an electromagnet, which operates automatically when a train reaches a certain point. This must be done as a collaborative exercise.

KEY STAGE 4 CONCEPTS

Pupils should be taught:

- *That a voltage is induced when a conductor cuts magnetic field lines and when the magnetic field through a coil changes.* The magnetic lines of force for two flat unlike poles are approximately parallel and straight between the faces. If the coil is turned, the wires at right angles to the field lines can be imagined as cutting them as they move. The cutting is essential to induce a voltage. When the coil itself is parallel to the pole faces, it is not cutting lines so no voltage is induced. As the coil rotates past this point, because its momentum carries it, the wires are moving the opposite way across the field so a reversed voltage is induced. The values of induced potential difference follow a sine curve if the rotation is steady. The cutting effect is similarly achieved if a magnet is moved instead of the coil. See Figure 4.2.
- *How simple A.C. generators and transformers work.* If the coil in Figure 4.2 forms part of a circuit then, provided a suitable way to connect the circuit can be found, A.C. is produced. The method for connection is called a slip rings arrangement (Figure 4.2).

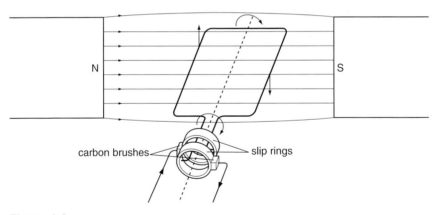

N   S

carbon brushes   slip rings

Figure 4.2

In the first diagram of Figure 4.3 the bar magnet's motion means that the coil is cut by the moving magnetic field lines. An alternative way of producing a changing magnetic field is to replace the bar magnet with an electromagnet running on A.C. Then the magnetic field would be continuously changing, lines continuously being cut and so an induced voltage continuously being created. The coil with an A.C. source is called the primary coil, that in which a current is induced is called the secondary. Both are wound on the same loop of iron, as shown in Figure 4.3, in the transformer.

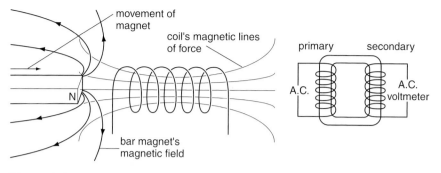

Figure 4.3

Since the induced voltage and current depend on the relative numbers of turns and resistance of the coils respectively, any desired values can be achieved. Transformers are widely used to step-down voltages for computers and door-bells and to step-up voltages to pass electricity efficiently to consumers. This is achieved through the National Grid. A low current/high voltage must be used. There is then the need for substations (transformers on a large scale) to reduce voltage for safe use by business and domestic consumers.

- *The quantitative relationship between the voltage across the coils in a transformer and the number of turns in them.* The voltage output of this secondary coil can have a value higher than that of the primary (step-up) or lower (step-down). The value is determined by the number of turns of wire which make up each coil. The relationship is simply:

$$\frac{V_p}{V_s} = \frac{N_p}{N_s}$$

where V_p and V_s are the voltages in the primary and secondary coils, and N_p and N_s the numbers of turns.

Example: How many turns are needed for the secondary coil of a transformer designed to provide 6 V for a toy train. The primary coil of 480 turns operates on mains voltage.

$$Mains\ voltage = 240\ V$$

$$\frac{240}{6} = \frac{480}{N_s}$$

So the number of turns in the secondary coil must be 12.

Such questions assume no energy loss in the transformer.
- *How electricity is generated and transmitted.* The National Grid is a network of

49

cables, mostly on pylons, linking generators and consumers through a series of transformers. A model is the best way to explain the need for transformers to pupils (see Activities).

KEY STAGE 4 ACTIVITIES

Inducing voltages

Review the magnetic field between the poles of a strong C-shaped magnet. The field lines run from the N-pole to the S-pole. Hold a piece of wire at its ends and move it up or down across the lines. You are cutting the lines with it. Repeat but with the wire connected to a centre zero galvanometer on a sensitive scale setting. (**Note:** these expensive instruments are easily damaged and must be set on *short* if moving them even a small distance). Treat with wonder the production of a 'no-battery' voltage.

- *How can we get a bigger voltage?* By moving the wire faster (other suggestions about the coil and magnets are welcome but you will not easily be able to vary them)
- *How can we change the direction on the induced voltage?* Move the opposite way.

You can then get a bigger voltage with a Westminster kit motor (or any other small motor with suitable connections). The motor is connected directly across the galvanometer and spun with the fingers, modelling a dynamo.

To extend the concept of 'cutting' magnetic lines of force, sketch a large drawing of the field between two flat poles on the board. Use a single piece of thick copper wire or make a giant coil with just one turn to keep it simple. Enlist a pupil or two to rotate the coil while you decide, with audience participation, when and where the lines will be cut. Clip two wires to the ends of the copper to show how they at once become tangled if you try to connect them to a meter. Two circles of copper, sellotaped to the terminals of the coil, can be used to explain the slip ring commutator and students should then realise that the induced voltage will be alternating. A generator is a larger version of this normally driven by steam.

From electromagnets to transformers

The steps suggested in the *Background* can be carried out using the Westminster kit. Iron cores are supplied with the kit which are easy to use and fit well together. Full instructions come with the kit and the induced currents can be demonstrated when moving a magnet relative to a coil, switching currents on and off, and using A.C. in the primary coil.

Transformers

A good large transformer makes an effective demonstration. Part of the iron loop is detachable for positioning of the coils. Use coils that are designed for mains voltage to show a step-down from 240 V to 12 V to light a mounted car head lamp bulb. If such coils are not available, it is nearly as effective to step down from 12 V to perhaps 1.5 V but darken the room for best results. The arrangement in Figure 4.4 will show no light on the 1.5 V side at first.

Having just one turn in the secondary coil means, using the formula given, only 0.5 V across the lamp. Adding a second turn will give 1 V so a small light may appear. Then a third turn will give correct brightness. If you can afford to dispense with the bulb add a further turn or two until it blows. If showing step-up function use only small voltages and use the formula to ensure you will not exceed safe values, around 30 V. Remember that only A.C. can be used in the primary and A.C. of the same frequency produced in the secondary.

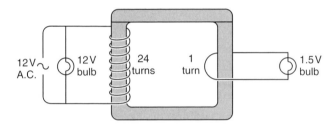

Figure 4.4

The puzzle of the jumping ring

In an arrangement like the one shown in Figure 4.5, switching on causes the aluminium ring to jump into the air and remain suspended. Pupils should be able to explain that the magnetic field of the coil induces a current in the aluminium ring. There is repulsion between its magnetic field and that of the

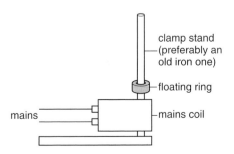

Figure 4.5

S coil, so the coil is suspended. Give the additional clue that the ring becomes hot; check it yourself before passing it to pupils. The coil must be a purchased mains coil.

You can vary this with a ring that has a split in it, hardly distinguishable from the complete one. When a pupil has a turn use the split ring (with a bit of sleight of hand). As it is not complete there will be no induced current and so no floating.

Quantitative work

When students use the formula they may find its representation in ratios difficult so it can be rewritten

$$V_pN_s = V_sN_p$$

which will prove easier particularly if simple whole numbers are used.

Distribution

In the countrywide distribution of electricity, the sequence is:

Power station	Transformer	Transmission	Transformer	Transformer
Generator	Step-up	Cables	Step-down to industrial consumer	Step-down to domestic consumer

The need for stepping up the voltage is to reduce the energy loss during transmission. Energy loss is related to the *square* of the current flowing so it is important to keep the current as small as possible. For the same power a small current is achieved by having a high voltage since *Power = Voltage × Current*.

S This is demonstrated by setting up model power lines and showing power loss

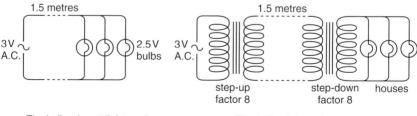

The bulbs do not light up due to a high energy loss

The bulbs light up because less energy has been lost

Figure 4.6

first without, and then with, the use of transformers. Do not attempt to do this, using mains electricity. Stick to the recommended voltages. Retort stands with insulating contacts for the wires represent pylons. See Figure 4.6. A 'Physics in Action' video called *Generation of Electricity* is available (Granada TV).

Assessing pupils' learning

- Imaginary sealed boxes can challenge students to say how they would find out what is inside (one or more magnets, magnetic materials, electromagnets with battery included). Rattling or opening the boxes is, of course, ruled out.
- Investigate the field patterns for a straight wire and make up a rule for deciding the direction of the field lines, a textbook will give the corkscrew rule.
- Some examples for using the transformer relationship should be given using relevant examples such as mains 240 V transformed to 6 V for a bell.
- Design a generator which uses wind power.
- Research into uses of induction e.g. electric guitars, slot machines, spark plugs.
- Competition to write a story containing the most electrical terms, puns encouraged. For instance, *Milli Watt* sent to the *cells* by *Sir Kit* for *vaulting* his *electric* fence.

ENRICHMENT AND EXTENSION ACTIVITIES

Students could consider whether there is interaction between wires carrying a current and design a way for testing their hypotheses – they may be given strips of cooking foil instead of pieces of wire as, being lightweight, they will flex under the small forces. Use a low-volt high-current supply in case the strips touch. An arrangement is shown in Figure 4.7, with the strips about 1 cm apart.

Two copper coils, if flexible, can be shown to repel or attract and this could be something else for the keen student to set up.

Greater detail of the construction of the transformer can be given to able students, details which improve the efficiency such as laminating the iron core and ensuring there are no air gaps in the core.

The relay can be investigated. It is a combination of an electromagnet with a reed switch. A small current flowing in the circuit containing the electromagnet can close the reed switch, which can be in a circuit carrying a large current so the device is effectively a switch.

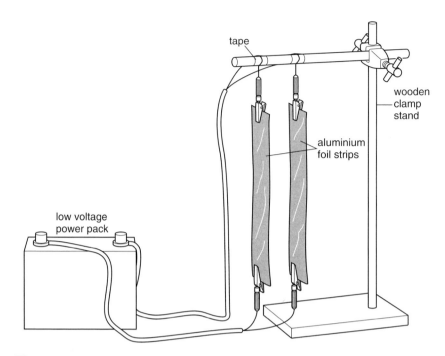

tape

wooden clamp stand

aluminium foil strips

low voltage power pack

Figure 4.7

Students can design relay circuits to:

- switch on lights when it gets dark (using a light sensor)
- switch on a freezer when it gets too warm (using a thermistor)

RESOURCES

Video
- Physics in Action *Generation of Electricity* (Granada TV).
- Scientific Eye *Electricity and Magnetism* from Yorkshire Television.

Information technology
- CD-ROM *Electricity and Magnetism* from Yorkshire International Thompson Multimedia Ltd Tel 0161 627 4469.

Teacher resources

- Nuffield Co-ordinated Sciences *Physics* for details of experiments with the Westminster kit.
- *Understanding Physics* by Robin Millar, published by Unwin Hyman, Section 30, for clear details of electromagnetic induction.

2 Forces and Motion

chapter 5

Linear motion and graphs

BACKGROUND

This section of the curriculum, which is about how and why movement takes place, is called dynamics. The principles governing the motion are Newton's laws. Successful at predicting motion for moving bodies on Earth and for planets and stars, they have only been superseded in this century by ideas of relativity and the quantum theory. These ideas are important only when considering very large objects or very small objects. For everyday terrestrial movement, Newton's laws function satisfactorily and are dealt with in the next two chapters. This one is a necessary precursor to understanding the laws, covering as it does speed, average speed and acceleration.

Ideas of faster and slower apply to most activities we carry out. Examples are abundant in sport, in games, in fun fairs and in simply getting about. Pupils in primary schools develop familiarity with the term **speed**.

At Key Stage 3 we want to show how we can measure the speed of a moving object in terms of the distance it travels in a certain time e.g. miles per hour, metres per second. The measurement of speed provides opportunities for practical work, which enhances pupils' understanding of the basic concept in which a rate of change of displacement is being measured. An understanding of the measurement and calculation of speed underpins so many topics in the sciences. Examples in different areas include:

- the speed of messages travelling along nerves

- rockets going to Mars
- ions moving through chemical solutions in electrolysis
- traffic flow
- shock waves such as in earthquakes

and countless others.

By using examples from different applications (see Key Stage 3 concepts) the appropriate units will be used and students will begin to link **rate** with **action per unit of time**.

Introduce older pupils to graphical representations of speed and displacement by showing them how to build up graphs from their own practical work. Rearranging equations to find either distance or time may still be difficult for average pupils so the triangle technique can be used here. In any case, numbers should stay simple and, as far as possible, whole. This is explained in the Activities below.

Common misunderstandings
- not distinguishing speed and velocity
- not distinguishing distance/time graphs and speed/time graphs

KEY STAGE 3 CONCEPTS

Pupils should be taught:

- *How to determine the speed of a moving object.* Two distinct pieces of information are needed in order to find out the speed of a moving object i.e. distance travelled and time taken to travel that distance. If the distance over a whole journey is used with the time taken to complete it, the *average* speed will be found. Use results generated by pupils own measurements as far as you can.

    Examples: Include a range of units in your examples such as kilometres per hour (speed-skiing), metres per second (100 m sprint), miles per hour (bus and train timetables)

    Note: 10 miles per hour is approximately 16 km/hour and $4\frac{1}{2}$ m/s.
- *The quantitative relationship between speed, distance and time.* Once you have discussed examples such as those suggested below and the pupils seem ready for a more formal treatment, give the equation in this form:

$$\text{speed} = \frac{\text{distance}}{\text{time}}$$

Go carefully through the units with the examples given but explain that for simplicity we usually keep to metres and seconds in science, giving us speed in metres per second, written m/s.

KEY STAGE 3 ACTIVITIES

Finding speed

You can introduce the measurement of speed through an activity which suits your group, the geography of your school and the equipment available. You can probably come up with other ideas but here are some suggestions. In each case ensure that pupils themselves say what two quantities must be measured and then how the speed is calculated. Encourage rounding up of measurements for simplicity of calculation and begin by questioning 'How far every second?' until pupils are confident with dividing distance by time to get metres per second (centimetres per second may be more appropriate for some activities).

- Measure the speed of pupils walking, walking backwards, hopping etc. in the lab or corridor or playground.
- With the agreement of the sports department go out to measure the speed of some athletes.
- Measure the speed of the traffic (big safety warning here – there must be a place where pupils can stand away from the road and where the distance along the road can be safely measured, and you must still go through the school's procedure for short trips outside the grounds).
- Make a 'Creeping Crawler' and have races with a follow up report (this toy is sometimes called by other names but good instructions with easy to find components are given in Resources).

Average speed

Suggest that a holiday journey to the seaside 100 miles away may take 2 hours, which clearly gives a speed of 50 miles *every* hour.

- *Can you do 50 miles per hour for the entire journey?* Clearly not e.g. traffic lights.
- *What have we found?* Average speed over the whole journey.
- *How is this possible?* You must have been travelling at a faster speed for some of the way.

ENRICHMENT AND EXTENSION ACTIVITIES

Design (don't carry out) the investigation of a snail's pace.

Ask pupils what is meant by the slogan 'Speed Kills'. Collect ideas for persuading drivers to slow down. Provide a plan of a town/village and give pupils the task of designing traffic calming measures. Write a leaflet for drivers about slowing down for pedestrians and cyclists, together with ideas for ensuring the leaflet reaches the right people.

Use a database such as Pinpoint to prepare a questionnaire on drivers' attitudes to speed, what causes accidents, what causes the worst injuries, speed limits etc. Use the database to process the results for the whole group and present the findings, perhaps in an assembly. For your reference approximately 10 under-15s are killed on the roads in Britain every week and many more injured. The peak time is between 4 and 6 pm on weekdays. At a speed of 40 mph or more, almost any impact with an unprotected person will result in death.

KEY STAGE 4 CONCEPTS

Pupils should be taught:

- *How distance, time and speed can be determined and represented graphically.* A graphical approach to this section is actually helpful and generally pupils do well with plotting straightforward graphs. Interpreting them is not quite so easy but again a practical approach (see Activities) gives you a way in and finding solutions from graphs can be easier than rearranging equations. The commonest error is to forget to look first at the type of graph you are dealing with. The choice is between a distance–time graph (a record of a journey) and a speed–time graph (a record of a speedometer reading).

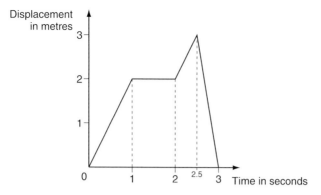

Figure 5.1 *Displacement-time graph.*

Figure 5.1 is a graph of displacement from a given starting point so, at this level, will almost always start at the origin, as shown. From there it is a matter of relating the graph to reality so in this example the sequence of events is:

1 Displacement increases (object moves away from start).

2 Displacement stays constant (object stationary).

3 Displacement increases (object moves even further away).

4 Displacement decreases (object turns and moves back towards the starting point).

From the graph we can find:

- the furthest distance travelled from the starting point (in this case 3 m) and total distance travelled (6 m)
- the time for each section of the journey (1 s, 1 s, 0.5 s, 0.5 s)
- the speed at any point in metres/second by calculating the gradient at that point (in this case the speed in the first section is 2/1 = 2 m/s and in the final section −3/0.5 = −6 m/s). The negative sign indicates travel *towards* the starting point

Guide more able students into realising that the straight line indicates a constant speed whereas a curved line shows a changing speed.

Figure 5.2 is a graph showing how speed changes with time. The sequence of events is:

1 Increasing speed for 1 s.

2 Constant speed for 1 s.

3 Increasing speed for 0.5 s.

4 Decreasing speed for 0.5 s.

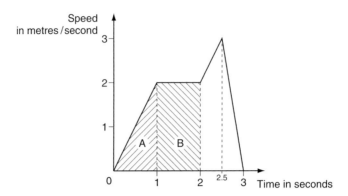

Figure 5.2 *Speed-time graph.*

From the graph we can find:

- The speed at any time (2 m/s after 1 s etc.).
- The rate at which the speed is changing i.e. the acceleration, by calculating the gradient of the graph (in the first section the change in speed is from zero to 2 m/s). A straight line shows a steady increase (or decrease) in speed, the same as **constant acceleration**. Note that here the negative gradient represents deceleration.
- Distance travelled during any or all of the journey. This is not simply **speed × time**, since speed is varying, but is given by the area under the line up to the required point. In our example suppose that we want to calculate the distance travelled in the first two seconds. The shaded sections show the area and we divide it into a triangle and a rectangle for easy working.

$$\begin{aligned} \text{Distance} &= \text{area A} + \text{area B} \\ &= (0.5 \times 2) + (1 \times 2) \text{ m} \\ &= 1 + 2 \text{ m} \\ &= 3 \text{ m} \end{aligned}$$

Care with units; the units of the axes are m/s and s so, when multiplied, give just m.

KEY STAGE 4 ACTIVITIES

Speed and graphs

The motion sensor
You can begin by just measuring the speed of someone walking across the lab. This raises issues such as:

- *When does timing start?*
- *Is the speed the same all the way?*
- *Do we all get the same answer?*

We can then reveal an excellent device called a motion sensor, linked to a computer. The unit sends out an ultrasonic wave (the waves used in baby scans) and then detects its return as it is reflected off anything in its path. The program provided calculates the distance of the reflected object and effectively draws a graph (distance vs time) of its movement. The object can be any pupil who can be persuaded to stand, walk or run up and back in front of the sensor. It's popular unless you get no volunteers, in which case you will have to move about yourself but remember to stay in line with the sensor.

You can draw out lots of discussion points here:

- *What is happening at this point on the graph?*
- *What does this straight line mean?*
- *What does this curved line mean?*
- *What does the gradient of the line mean?*

Once speed has been suggested show that the program can calculate this and will plot it over the original graph. It will also go as far as finding acceleration for any groups ready to see this, or you may reserve it for another occasion. A bonus is a series of preset graphs in the program. Challenge pupils to try and match the graph on the screen by moving appropriately. You will usually get plenty of volunteers. If your school has not got a motion sensor do put in an early request – they won't regret it.

Trolleys

Dynamics trolleys have traditionally been used to link speed and graph work. They should be seen but definitely not overused as less tedious methods make their point more snappily. In essence they are wooden trolleys with low friction bearings whose speed can be measured by pulling a piece of tape through a device called a ticker-timer. The timer will place a dot on a specially made strip of paper every 1/50 th of a second (mains frequency). If the paper is stationary all the dots will be on top of one another but when pulled through there are spaces between the dots. The further the spacing the greater the speed of the trolley. Since the time interval is known to be 1/50 th of a second the speed can be calculated using the distance between the dots. However the small distances and the difficulty of dividing by 1/50 th do not make this practicable for many pupils. The design enables pupils to build up a graph of the movement of the trolley without the need for real time calculations. The procedure is:

1 When you have made your strip mark every tenth dot (10 *gaps*) by drawing a line across the strip.

2 Number each 10-dot strip consecutively (in case any one sneezes).

3 Cut across on the marked lines so you will end up with lots of strips.

4 Stick them side by side on a base line in order.

I usually start this activity by letting pupils walk while pulling about a metre strip through and then ask what the graph says about their walk – in Figure 5.3 getting faster then getting slower. Compare with friends.

Strips containing the same number of dots measure identical time intervals, so Figure 5.3 is effectively a graph of speed against time. Pupils can make a different graph using the trolleys on the flat, smooth tracks usually provided. Give a warning about the delicate bearings, trolleys should not be allowed to fall. Releasing a trolley down a sloping track supported at one end on a few books is a good first investigation; suggest stopping the trolley at the bottom with a rolled up coat. Support the ticker-timer on the same books and stick the paper to the trolley.

The result will be steadily lengthening strips and pupils will appreciate that this indicates a steady increase in speed i.e. constant acceleration. To use the strips to calculate average speed for a journey find the length in centimetres of 5 consecutive 10-dot strips. This represents one complete second and so the average speed in cm/s has been found. The trolleys' acceleration *can* be found but is best left without units simply as the gradient of the graph in Figure 5.3, the steeper the graph the greater the acceleration. Since 10-dot strips are always used this enables accelerations to be compared with minimum of calculation.

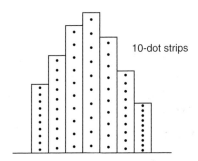

10-dot strips

Figure 5.3

Try to encourage pupils to feel at home with the idea of calculating not only speed but distance and time by slipping in a few mental arithmetic examples.

Example: If I push my supermarket trolley at 2 m/s to a checkout 20 m away how long will it take me? Will I beat Mr Jones who takes 12 s?

For less simple problems offer the triangle technique.

Triangle technique

This consists of placing the three elements of equations of this type into a triangle which pupils seem to both learn and operate more easily than the equation.

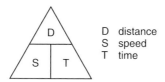

D distance
S speed
T time

Figure 5.4 *The triangle technique.*

The technique is to cover with the finger the item to be calculated, then the position of the remaining two items shows how to do the calculation. For instance, covering *t* leaves *d*/*s* whereas covering *d* leaves *s* × *t*. Give it a try with other equations such as *density = mass/volume* and *power = voltage × current*. It's crucial of course to get the letters in the right place.

To solve problems of speed etc. graphically, pupils can begin by plotting graphs of simple journeys, e.g. a pizza delivery person, a plane etc. and making calculations as detailed in the concepts above. So far we have used m/s as our unit of speed but more able pupils should be introduced to m s^{-1} as a unit.

Assessing pupils' learning

- You will find lots of examples in standard texts using graphs to solve problems about motion, see resources.
- Ask pupils to describe how they might use a motion sensor to find the speed of cars through a residential area. Expect graphs with figures to suggest the results and ask for any problems which might lead to inaccuracy e.g. difficulty of aiming as sound waves spread out, effect of air density changes when hot, damp, etc.

ENRICHMENT AND EXTENSION ACTIVITIES

If you have the facility some multiflash photography is interesting to do. Your essentials are a room with good blackout, a good strobe lamp and a camera with

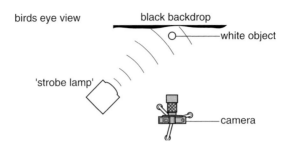

birds eye view   black backdrop

white object

'strobe lamp'

camera

Figure 5.5

a B setting set on a tripod. Careful lining up is required and may need to be done before the lesson.

Use a simple movement by a light coloured object such as a covered table tennis bat striking a ball, a large pendulum or a falling white ball. Include a vertical rule marked with a white strip every 10 cm. A fast black and white film will be needed. Be aware that you won't have the pictures available in the same lesson unless they can be done in a small developing tank. *WARNING:* flashing lights can trigger epilepsy – they should be used above 15 flashes per second, for this experiment 20 to 30 per second works well. The negatives can be projected onto a screen and, with the help of the rule, measurements made.

The term *velocity* will crop up when pupils tackle more advanced questions and you should avoid using it interchangeably with speed unless you specify that the motion is all taking place in one direction. Normally speed (just a numerical value) is what is required but velocity must be used if we are considering speed *in a certain direction.* It is referred to as a vector quantity (see Chapter 7 for an example). An object, for instance the Moon, can travel at constant speed but have a changing velocity because its direction is continually changing.

More able pupils should handle problems involving moving around a circle e.g. how fast is the surface of the earth at the equator travelling around the axis?

$$\frac{\text{circumference}}{24 \text{ hrs}} = \frac{40\ 000\ 000 \text{ m}}{86\ 400 \text{ s}}$$
$$= 465 \text{ m/s}$$

How fast does an hour hand/second hand move around your watch face given its radius? (Analogue watch owners to supply data.)

A possible investigation here is to look at the factors affecting the speed of an object sliding down a slope. Variables include the height of the slope, area of contact, length of slope and types of material.

RESOURCES

Video

- Scientific Eye *Force and Friction* from Yorkshire Television.

Pupil activity

- *Fun with Science* by Brenda Walpole, published by Kingfisher 1991 ISBN 0-86272-241-1 includes details of the Creeping Crawler.

Teachers resources

- Examples of interesting speed calculations for a range of pupils are to found in Co-ordinated Science *Physics* by Stephen Pople and Peter Whitehead, Section 1.
- For some harder problems see *Physics Matters* by Nick England, Sections 1–3.

chapter 6

Balanced and unbalanced forces

BACKGROUND

Classical theories of motion were based on the assumption that being stationary was the natural state of a body. Bodies moved in order to achieve that position, heavy bodies downwards, light bodies upwards. To achieve motion towards an unnatural state required an external agency. Circular motion was not linked directly with linear motion. Such ideas seemed to fit everyday experience and were accepted until careful experimental work revealed conclusions which were inconsistent with them.

Galileo performed important experiments in dynamics, see Key Stage 3 Activities, paving the way for Newton. The novelty of his approach was in considering that motion does not require an external agency but only *changes* in the movement. Changes in speed include starting from rest, accelerating, changing direction, decelerating and stopping.

There are conceptual difficulties with this that can persist, even in Physics students at University. They arise because the theories about motion put forward by Newton appear to conflict with everyday experience. Pupils' own perceptions are continually being reinforced whereas the 'Laws of Motion' seem to be something alternative, which happens in the laboratory or outer space. An understanding of the nature of the external agency, force, is crucial. This is the basis of this chapter.

A change in speed or direction is always caused by a force. There may be several forces acting on an object but any change in speed or direction is due to the resultant force. On Earth there is always gravity acting, pulling objects down and there is always friction. Hence the need to use the laboratory, where we can try to control these, or to refer to outer space to provide evidence for our arguments.

When an object *slows down* it's because the resultant force on the object acts in the *opposite* direction to the movement.

> Example: *A bungey jumper reaches the bottom of the jump, the tension in the cord is the force which slows the victim down.*

> Example: *A swimmer glides through water but is slowed by the force of friction and will need to take another stroke.*

When an object *speeds up* it's because the resultant force on the object acts in the *same* direction as the movement or moves it from rest.

> Example: *When a water skier is pulled behind a launch, the tension in the line speeds up the skier, though not indefinitely.*

> Example: *If a table tennis ball is falling, the resultant force is the weight of the ball due to gravity minus the small force due to the friction of the air.*

We have used the term **resultant force** here because there will probably be more than one force acting. In some cases they might balance to give a resultant force of zero. If so there would be no change in speed. If the speed is 0 m/s that is easy to accept but if the speed is, say, 3 m/s in a straight line then surely there must be a force keeping the object going? The answer is no, no force is needed unless there is an opposing force, like friction, to overcome.

Common misunderstandings

- if something is moving there is a force acting on it
- if something is not moving there is no force acting on it
- a moving object stops when its force is used up
- a constant force means a constant speed

KEY STAGE 3 CONCEPTS

Pupils should be taught:

- *Ways in which frictional forces, including air resistance, affect motion e.g. the effect of air resistance on a descending parachute, the effect of friction between a tyre and a road.* Friction both hinders and helps us. We could not move without friction between tyres and the road or foot and path, nor would our brakes work without it. On the downside if our aim is speed, and low wear and tear, friction acts against us. Whenever you want to judge whether friction is helping or hindering ask yourself how the activity would work on wet ice or with lubricated surfaces. Pupils at Key Stage 2 will have ideas about air resistance and will probably have investigated parachutes and paper darts. As with all work on forces, diagrams showing direction should always be used (see Figure 6.1). Be careful with objects slowing down. For the parachute the driving force is the weight which stays constant. For the car, if brakes are being

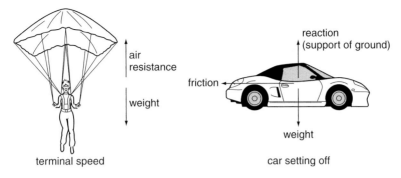

Figure 6.1

applied there will be no driving force so only one horizontal force is shown on
the diagram.

The whole picture of the parachutists' movement is dealt with in Chapter 7.
Some pupils may feel that the car must come to a dead halt if there is no
driving force. Use the idea that a moving body has energy. Until this energy is
transferred the car continues to move forward.

Note that the retarding force in air or other fluid, known as air resistance or
drag, becomes greater
- as the speed of the object becomes greater
- as the cross sectional area facing the direction in which the object is
  moving increases

The situation between solid surfaces is best treated separately from fluids. It is
usual to think about a close-up view of the surface as very uneven. When the
surfaces are dragged in opposite directions vertical bits of the surfaces are not
easily able to pass one another. It is like dragging pieces of coarse sandpaper
over one another. If a surface is very smooth there are very few vertical faces.
An example is PTFE (non-stick pan coating). The friction between two
particular solids is dependent on the weight of the upper object and on the
relative speed of the two surfaces.

- *That unbalanced forces change the speed and/or direction of moving objects.* For
  instance, an unbalanced force is provided when you:
  1 press the accelerator pedal of a car

  2 push down on the pedal of a bike

  3 catch the wind in a surfers sail

  4 get pulled up in a lift

As you begin to move in the direction of the unbalanced force there is very
little opposing force because you are moving so slowly. As you speed up, these

opposing forces get bigger until they balance the driving force (thrust), setting a top speed. The opposing forces are as follows:

1 Friction between the moving parts of the car, air resistance and some tyre friction, particularly in the case of soft tyres.

2 as 1.

3 Water and air resistance.

4 In this case, it is more likely that the driving force will reduce since it is not desirable to have a lift accelerating for more than a short time. The opposing force is the force due to gravity but that is constant (see Chapter 7).

In each case acceleration continues only as long as there is a resultant force in any direction.

- *That balanced forces produce no change in the movement of an object.* In each case in the preceding paragraph as soon as the point is reached where the forces are balanced the speed ceases to change and the vehicle continues at the same speed until there is some other change of force.

1 Brakes are applied creating extra friction.

2 The same.

3 The sail is turned producing a different driving force.

4 Motor speed reduced still further.

- *That forces can cause objects to turn about a pivot.* The word pivot needs explaining; you can begin by using turning point. It is important to understand its meaning as calculations follow where distances must be measured from this point. The turning effect of a force, the moment, depends on the size of the force and how far it acts from the pivot.

---

moment of a force = force × distance from pivot

$$N\ m\ =\ N\times m\ \text{or}$$
$$N\ cm\ =\ N\times cm.$$

Example: What is the moment of a force of 500 N acting on a pedal 20 cm from an axle?

$$moment\ of\ force = 500\times20\ N\ cm$$
$$= 10\ 000\ N\ cm$$

- *The principle of moments and its application to situations involving one pivot.*

For a system in equilibrium the principle of moments states that the sum
of the clockwise moments equals the sum of the anticlockwise moments.

More simply 'if a pivoted lever is balanced then the left hand moment equals
the right hand moment' will be sufficient to deal with the problems
encountered at this level.

*Example: A see-saw, with the pivot in the middle, is balanced when one person
sits on each side. The adult weighing 500 N sits 2 m from the pivot. Where must
the child weighing 250 N be sitting?*

$$Left\ hand\ moment = 500 \times 2\ N\ m$$
$$Right\ hand\ moment = 250 \times d\ N\ m$$
$$500 \times 2 = 250 \times d$$
$$4 = d$$

The child is sitting 4 m from the pivot.

KEY STAGE 3 ACTIVITIES

Parachutes

If pupils have not had the opportunity to make and test parachutes they should
have a chance now. Washers as heads with paper bodies make satisfactory people
and show a degree of crumple when they land. Parachutes can be made from
pieces of material. Pupils can decide for themselves what shape they want the
envelope to be, and how to ensure a fair test. The possible variables include
weight of material, area of envelope, length of string with time of fall being
measured. If time permits paper darts are best as an end of the lesson activity, the
winner being judged on either distance travelled or time of flight or one of each.
Ask pupils to say what their aim was in deciding the shape of their parachute or
dart and encourage ideas about air resistance.

To investigate its effects horizontally, attach a sheet to a vehicle. It would be
confusing to call it a sail, which aids movement, though it will look like one. The
vehicle (car or trolley) can be allowed to roll down a gentle slope and timed
without its resistance sheet and then with it attached (a lolly stick and card will
do). The area of the sheet can be varied. If so start with the largest sheet first to
avoid using vast quantities of card.

Pupils who have difficulty with area and graph construction can experiment
with folding the card into different shapes to find how the speed is changed.
Alternatively a spreadsheet would be useful here – see Resources.

Falling shapes

Ideas about friction and streamlining can be gained from investigating differently shaped objects falling through tubes of viscous liquid such as glycerol. It is sometimes suggested that pupils can make shapes from plasticine to drop, but the plasticine seems to suffer horribly as a result and is very unpopular with anyone clearing up.

It would be better to assemble some shapes in metal or plastic and keep them for this experiment. You could use ball bearings, nails, cubes, old biro cases, coins etc. Look out for objects (bullets!) which are symmetrical about the axis pointing the way they will fall. They should already be in a dish of the liquid beforehand so that air bubbles don't cause odd readings and warn pupils that, once dropped they can't easily be retrieved. It's possible to draw the magnetic ones up with a magnet but usually requires more patience and time than is available.

Some pupils will realise that a fair test would demand all be the same mass. They should be credited for good thinking and allowed to wrap some wire around lighter objects to achieve approximately the same mass. Include extra paper towels in your apparatus order!

Hovercraft

The hovercraft as a friction free vehicle can be investigated with a model using a balloon on a base. The lightweight base, which can be shaped like a hovercraft, has a small hole in the top. The mouth of a blown up balloon is pushed into the hole so that the air is forced out downwards lifting up the craft. This can be seen to float at roughly constant speed until the air runs out. Try to get a run so you can see the effect of no unbalanced forces, blowing from in front, blowing from the side using straws. Try to get circular movement by running round the table, or get a volunteer. Notice that you blow *towards* the centre of the circle.

Galileo

The arrangement in Figure 6.2 was a thought experiment by Galileo who forwarded ideas in mechanics so much. Two to three pieces of shaped track are needed, preferably made of a polished metal to reduce friction e.g. a curtain track. A steel ball bearing is used to roll down the track. Demonstrate from a long run of bench.

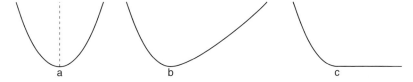

Figure 6.2

a) The ball will roll back and forth and should rise to a similar height on each side. It won't quite, of course, but you can explain that there is bound to be some loss of energy as the ball rolls.

b) The ball rises but does not quite fall off the other side.

c) The ball will not only leave the track but usually race straight to the far wall with a satisfying smack. This is a ball moving in a straight line without speeding up or slowing down (much). It's usually an investigation which is remembered so you can refer to it when pupils 'forget' this law.

- *Will the ball roll for ever?* No, it will meet resistance or friction.
- *Where could we go to get away from friction?* Discuss all suggestions. They will probably include ice (so apply the principle to a skater on an endless pond), maybe a hovercraft (someone who has been on one can describe it to everyone else), and space (apply to a rocket, no fuel is needed once the Earth has been left behind).

Sliding

Since forces are given in newtons pupils should try an investigation where they can get actual values for frictional forces. They can do this by sliding objects along surfaces pulling with a forcemeter. Demonstrate, by sliding a shoe along a suitable surface, that the reading gets bigger until the shoe starts to move and then suddenly gets smaller again. By pulling steadily as the shoe slides a constant reading can be obtained and this is the reading to obtain. Note: **limiting friction** is the biggest reading just before the block begins to slide and **sliding friction** is the constant reading as the block slides steadily along.

Pupils can then find out who has the shoe least likely to slip and/or which is the best surface to slide on. A range of surfaces can be provided e.g. a carpet tile, piece of vinyl, wood, plastic, wet surface (in a tray) and ice. Remember to get a layer in a tray frozen beforehand. Ice is not very slippery when dry but covered with a layer of water it is very different. The pressure cars apply melts ice or snow on the surface and the subsequent reduction of friction makes it extremely dangerous.

If pupils select one surface and measure the sliding friction between it and a rectangular block they can so some quantitative work by turning the block onto its other surface and adding different weights to it.

Turning forces

How can we turn some of these as easily as possible: steel nut, tin opener, car jack handle, heavy door?

Provide some simple drawings for pupils to add an arrow to show where and in what direction the force is applied. They will begin to realise that it is always as far from the turning point as possible. Ensure that this is called the pivot. They can investigate the relationship between force and distance from the pivot as follows: a half metre rule can be attached at one end of the bench by using a piece of tape to make a hinge. The other end can be raised using a forcemeter so that the force and distance from the pivot can be read. It will quickly be seen that the smaller the distance the greater the force – in fact double the effort at half the distance.

An interesting application of this is how bones act as levers, for instance, the lower jaw. Pupils can feel which muscles are being used when they chew and where the pivot is. Why are the molars used for chewing, situated at the back of the mouth (nearest the pivot)? Lots of examples of levers will be shown in your textbook.

For balancing moments use a real seesaw and some bathroom scales or a ruler and some small weights. All that needs to be shown is that a weight on one side will only balance a weight on the other side if the two moments are equal. Some schools have square pieces of metal, usually brass. These do work well but you need to tell pupils to place them on the ruler or wooden beam diagonally across otherwise it is difficult to judge the exact distance of their centre from the pivot.

Rules for success are:

- balance the beam at its centre, adjusting if necessary with small pieces of plasticine
- only allow one mass or coin on the 'fixed' side and tape or blu-tack it to the furthest position
- allow any number of masses or squares on the other side but they must be in a pile, not scattered

Ultimately the pattern will arise that doubling the distance halves the mass required. The statement of the principle required by the National Curriculum is a bit daunting so follow it with lots of simple examples of objects on seesaws and ideas of weighing objects using the principle. Don't expect pupils to be adept at dividing through to find a weight, stick to distances of one or ten units, whether using metres or centimetres.

Assessing pupils' learning

- Ask pupils to imagine they are technicians who have to test designs for the best shoe for abseiling – perhaps an abseiler in the school could explain to pupils what is desirable in such a shoe. This could be carried out as a full investigation.

- Model limbs can be made using thin dowel or stiff card for bones and rubber bands for muscles/tendons. Explain that the tendon joins the muscle providing the force to lift the bone. They can feel the tendon from their bicep in the crook of their elbow.
- Design and make a balance to weigh a coin using the principle of moments. They can compare the weights of different value coins.

ENRICHMENT AND EXTENSION ACTIVITIES

Making balancing toys is a nice activity that helps some pupils to understand the idea of equilibrium and moments. See Resources below.

Research into shapes of racing cars and fish and why they are different.

The manually dextrous can make a balance capable of weighing something very small such as a seed or a fibre. Details can be found in older textbooks such as the original Combined Science course for 11 to 13 year olds. In the design a straw is pivoted on a pin, a counterweight loop of fine wire sits on one end and the object to be weighed on the other. Balance can be achieved by sliding the loop of wire along extremely carefully. The weights can just be comparative or the balance can be calibrated using small pieces of graph paper cut from a whole sheet, which has been weighed.

Have a parachute competition if you have a balcony or landing to drop them from.

Faster moving pupils might look at the Bernoulli effect, one which is responsible for providing uplift on aeroplanes. They hang two pieces of paper about 3 cm apart and blow through the space between them. You might expect them to move apart but the reverse is true. Where the pressure of air is less, between the pieces, they are sucked together. Pupils can investigate this effect by blowing across wing shaped structures to show the upthrust.

KEY STAGE 4 CONCEPTS

Pupils should be taught:

- *About factors affecting vehicle stopping distances.* Stopping distances are made up of two parts, the thinking distance and the braking distance. The thinking distance is the distance travelled during the driver's reaction time i.e. from awareness to applying the brake. It depends on the speed of the vehicle as well as the natural reaction time. The braking distance is the distance travelled during braking and it depends on the speed of the vehicle and the force applied. For a constant braking force the distance travelled depends on the square of the time taken to come to rest. For two similar vehicles, B travelling twice as fast as A, the braking distance will be four times as far for B as for A. Triple the speed gives nine times the braking distance and so on.

The two distances added together give the total minimum stopping distance, a frighteningly large distance. Pupils will readily be able to think of ways in which these might be increased. Thinking time can be increased by loud music, distraction, tiredness, medication, intoxication, drugs and low concentration (maybe a personal crisis). Braking time can be increased by poor brakes, wet roads, icy conditions and bald tyres in wet conditions. Poor visibility would not actually increase either of these times but would increase the time before awareness.

- *The difference between speed and velocity.* There are quantities called vectors that are only fully specified by both size and direction.

   Examples: displacement, velocity, acceleration, force, momentum.

   Quantities which do not have direction associated with them are called scalars.

   Examples: mass, temperature, distance, speed.

   Vector quantities can not be added as simply as scalars because of the need to account for direction. To illustrate the difference, compare these two journeys (Figure 6.3). In a) a boat is rowed at 3 m/s due N, in b) the same boat is rowed at the same speed but, at the same time, a current carries it 3 m/s due E.

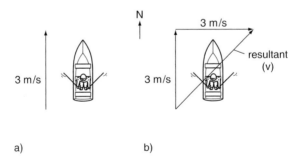

a)   b)

Figure 6.3

To find the true speed of the boat relative to the bank we have to do a vector addition, taking directions into account. We need to find a **resultant** velocity, i.e. one which you could use to achieve the same result as the two component vectors. To find the resultant:

Either

1 Make an accurate scale drawing of the two vectors and their directions. Measure the side marked resultant.

or

2 Complete the triangle and use Pythagoras' theorem to solve.

$$v^2 = 3^2 + 3^2$$
$$= 9 + 9$$
$$= 18$$
$$v = 4.2 \text{ m/s North East}$$

The resultant velocity is 4.2 m/s 45° East of North.

Similar calculations can be performed on any pairs of vector quantities provided the triangles are correctly constructed. The two components should follow one another but with their directions correct. The resultant V should point the opposite way around the triangle. The calculations are possible, though more tricky, if angles other than 90° are involved but able pupils can apply trigonometry. For others, solution by careful drawing is always possible.

Note: If the resultant and one component are given, similar methods can be used to find the other component.

• *About acceleration as change in velocity per unit time.* The acceleration is the rate at which velocity is changing. Since velocity is the rate at which displacement is changing this means that it is a rate of a rate – the units reflect this with the s² term. Put another way, a Jaguar can go from 0 to 60 mph in 6 seconds. That means it changes by 10 mph every second. That could be expressed as 'acceleration is 10 miles per hour per second'.

In standard units:

$$\text{acceleration} = \frac{\text{change in velocity (m/s)}}{\text{time taken for the change (s)}}$$

These units can be written metres per second per second or m/s² or ms⁻²

KEY STAGE 4 ACTIVITIES

Stopping distances

This topic can not easily be taught through practical investigation unless you can afford to lose a few of your pupils! The best alternative is to look at the highway code section giving stopping distances. Measure out the stopping distance for, say 50 mph. It is surprising how general the belief is that cars stop dead. Discuss contributing factors to these times (see Concepts). Put ideas about extenuating

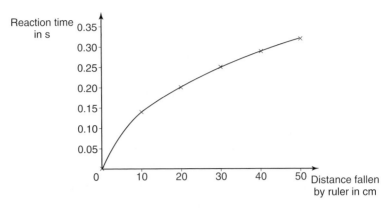

Figure 6.4 *Conversion graph for reaction times.*

circumstances on posters for feedback. More able pupils should be able to see a general relationship between speed and braking distance, others just need to know that if speed is doubled, braking distance is four times as big.

Reaction times can be measured although this may well be covered in biology. The method is for one pupil to sit with a hand held around but not touching a half metre ruler which is held at the top by a partner. Without warning the partner drops the ruler and the first pupil responds by catching it. The distance fallen depends on the pupil's reaction time; the longer the reaction time the further it falls. A conversion graph can be given so that the reaction time can be worked out (see Figure 6.4).

It's worth pupils finding out that they get better if they practice but worse if they are not concentrating.

Vectors

To establish the idea of displacement as a vector have pupils walking around the room in paths similar to those shown in Figure 6.5.

In each case measure and then calculate the total displacement, D, from O, the starting point. It *does* matter which way you are going. Explain that displacement D is a resultant, that is, it could replace the two separate

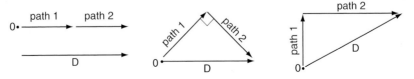

Figure 6.5

components. Show how you can find the resultant by drawing or, with appropriate groups, by calculation.

This will form an introduction to the idea of velocity as the vector correspondent of speed. If an object is subject to two velocities at the same time, such as a dinghy crossing a river (see Concepts) then what single resultant velocity could replace them?

The pitfall is to confuse the picture of what is happening with the vector diagram, and this seems to be even more difficult when calculating resultant forces. As good practice ensure that pupils always draw two diagrams.

Whichever method of solving you use choose very simple numerical examples. Getting bogged down with 'hard sums' is one reason for the perceived difficulty of physics. Always allow the option of solving by construction as equally valid.

Acceleration

- Introduce acceleration through a range of advertisements for cars which state '0–60 mph in 6 seconds' etc.
- Analyse in groups what that means i.e. a change of speed from 0 mph to 60 mph takes 6 seconds.
- If the change in speed is steady, what would the speed be after 1 second, 2 seconds? Every second it changes by 10 mph.
- We can say that the acceleration is 10 mph per second. i.e. 10 miles/hour/second
- Work through a similar idea using metres so that the units are the more commonly used m/s^2.

Example: A runner accelerates steadily from rest to a top speed of 8 m/s in 4 seconds. What is their acceleration?

$$acceleration = change\ in\ speed/time\ taken$$
$$= 8/4$$
$$acceleration = 2\ m/s^2$$
This means that every second, speed increases by 2 m/s.

The use of a **linear air track** and **light gates** provide a good demonstration at this point.

The air track is a long narrow tent-shaped tube with a lot of holes in the two slanting sides. When the end is connected to a vacuum cleaner set to blow, air is forced out through the holes to make little jets. Let pupils feel this before going any further.

There are specially designed vehicles which sit on the track and can hover on the air jets. When given a small push they glide along almost friction-free. As the push of the air is upwards against gravity we have effectively removed our problem forces, weight and friction, so this opens up greater opportunities for reliable motion experiments. The track should be carefully levelled; if perfect the vehicle should remain at rest if no horizontal force is applied and should glide at constant speed given a slight push. Get it as close as you can. Make sure the ends of the track have rubber bands stretched across so the vehicle will bounce off and glide back again.

Let pupils see what happens when the air jets are switched off (but it's not equipment for volunteers to help with as it's very important for the track to remain really smooth). Allow the vehicle to glide to and fro while you ask:

- *What is happening to the speed?* Not much along the track but slower after each rebound.
- *Why?* Energy transferred to the rubber bands.
- *What evidence is there?* The rubber will get hot, the vehicle will stop moving.
- *What if we had a really, really long track?* It would go on indefinitely since there are no unbalanced forces. There is no driving force and no friction force.
- *What would improve the investigation?* Being able to measure the time accurately.

This is where you can introduce the light gates. The gate is a device which contains a light source (or infra red like the remote control for TVs) and a sensor (photodiode). Each is linked to a computer with a suitable program that measures the time when the light beam is being broken. If a computer and program are not available the gate can be linked to a millisecond timer. Show pupils that only time is measured, by passing objects through the gate. The metal gates are quite robust so as long as you hang on to them.

Then arrange two light gates in clamps over the track so that the vehicle passes through them (usually a 10 cm piece of card is attached to cut the beam). Show how the time is measured. The program offers the opportunity to calculate the speed (or velocity since all movement is in one line) and will ask you to enter the width of the card. You may be able to choose to have the calculation in easy stages or just an answer.

Attach a *small* weight to the vehicle on a long string and arrange for it to fall over a pulley wheel. The vehicle will be seen to be getting faster. Ask pupils how the acceleration can be measured. Speed must be measured in two places and the time to travel between the two measured as well. The program will be able to do this with ease and provides a good example of how scientists can make use of IT for eliminating tedious and repetitive calculation. Some pupils might be able to suggest how a spread sheet can be prepared to do this and how to test it by substituting values for t (time).

An effective way to deal with the presence of gravity and friction in our experiments is to do virtual experiments using a CD-ROM, see Resources. Resources also includes a video as part of the *Science in Action* series which tackles the difficulties associated with 'balanced forces'.

Balanced and unbalanced forces

Assessing pupils' learning

- An investigation is possible here using toy cars running (not skidding) down a slope. The distance they travel can be measured and compared with the angle of the slope or the position on the slope from which it is released. The variables are too limited for a GCSE Sc1 but it would form a useful practice in the early part of the course.
- Slower pupils will need more time with ideas about changing velocity but it is worth persevering with acceleration even if the units are memorised as it will be needed for work on gravity.
- Clearly practice will be needed in handling vectors and acceleration calculations. Examples appropriate to ability will be found in almost every physics textbook for this age range.

ENRICHMENT AND EXTENSION ACTIVITIES

83

Pupils can be given tasks associated with road safety, from posters against drinking and driving to reports recommending action to the government to reduce speed on roads.

Community policemen are usually willing to come and talk, provide some scary statistics and judge the poster or related work. I remember being given a drink of whisky by a speaker to show how it changed my reaction time but that was a long time ago!

Work on vectors, other than being able to distinguish between speed and velocity, should be targeted at those with good mathematical ability. Additional explanations are given in the reference below together with worked examples and an interesting report on tachographs.

A good investigation involves the autogyro, usually called paper helicopter. It is an open-ended experiment with the possibility for everyone to discover something. Figure 6.6 shows a possible design and recommended variables are:

- height of drop
- material
- length of wings
- width of wings
- mass (number of paper clips)

Most able students will realise that some of these are interdependent and think carefully about how to maintain a fair test. All will enjoy it.

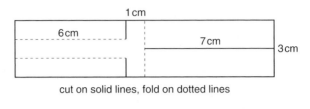

cut on solid lines, fold on dotted lines

attach paper clip

Figure 6.6 *The paper helicopter.*

RESOURCES

Information technology
- Excel is a spreadsheet that could be used for the extension activity suggested.

Video
- Science in Action series entitled *Forces and Pressure.*
- Scientific Eye *Speed and safety.*

Teacher resources
- *Fun with Science* by Brenda Walpole, published by Kingfisher.
- Vector work is usefully dealt with in Sections 5.5 to 5.8 of *Understanding Physics* by Robin Millar published by Unwin Hyman.
- For help with using a spreadsheet see *School Science Review*, Dec 1997, Vol 79 (287) Page 69.
- A useful CD-ROM called *Motion* from Cambridge Science Media, 354 Mill Road, Cambridge CB1 3NN. Tel 01223 357546.

chapter 7

Newton's laws of motion and gravity

BACKGROUND

This topic builds on the ideas of forces and their effect on motion, which was introduced in Chapter 6.

Newton's laws were a long-lasting contribution to an understanding of dynamics. They achieved links between earthly motion and that of heavenly bodies. The effect of force on motion was replaced by a consideration of its effect on a *change* in motion. This important difference enabled progress in linking change in velocity and force. Be sure to stress that velocity is a vector quantity, that is it can be said to have changed if either its magnitude or direction or both change.

A resultant force at an angle to the direction of movement of an object causes a change in direction. If the force is at right angles to the movement, such as in throwing the hammer, the object follows a circular path or orbit. Newton was able to identify the force that pulls objects to Earth with that which keeps planets in their orbits – namely gravity.

The force of gravity exists between all objects that have mass but the forces between the masses we encounter daily are too small for us to measure or notice. The force between a body and Earth is noticeable because it makes objects, including us, fall to the ground. The effect on the Earth is negligible.

Common misunderstandings
- acceleration of falling masses depends on their mass
- there is no gravity on the Moon
- gravity is a property of air pressure, so there is no gravity once you leave the Earth's atmosphere
- objects are pulled by the Earth's gravity, they do not themselves exert a force
- you do not need a force to make things fall, 'they fall naturally'
- gravity acts in various directions depending on the situation

KEY STAGE 4 CONCEPTS

Pupils should be taught:

- *That balanced forces do not alter the velocity of a moving object.* For an object at rest, the downward force due to gravity is exactly balanced by the upward force provided by the reaction of the support. For instance if you are relaxing in your seat the downward force is your weight in newtons and the upward force is the reaction provided by the chair. If your chair is particularly comfortable then the force is provided by the tension in the elastic parts of the seat. If the seat is hard we simply call it the reaction, though it must also be provided by a change in shape of the seat. Similarly there is a reaction on the chair legs by the floor. In each of these, forces are balanced and so there is no change in velocity. Since the velocity happens to be zero the items concerned remain at rest.

  For a moving object with no forces acting on it, such as a spaceship far away from any planets or stars, the movement continues indefinitely at the same speed and direction. For vehicles on Earth, travelling with constant velocity, there will generally be four forces acting, thrust (forward), friction (opposite to thrust), weight (vertically downwards) and reaction (opposite to weight). For experiments to agree with the Laws of Newtonian mechanics we must take this into consideration.

- *The quantitative relationship between force, mass and acceleration.* This relationship is one way of stating Newton's Second Law. The relationship is:

$$\textbf{force} = \textbf{mass} \times \textbf{acceleration}$$

In terms of units:

$$N = kg \times m/s^2$$

We can write this as:

$$\text{force} = \text{mass} \times \frac{\text{change in velocity}}{\text{time interval taken}}$$

Mathematically

$$F = \frac{\delta(m \times v)}{\delta t}$$

where δ means a change in the quantity written next to it.
m can be included with v since it is constant.

$$F \times \delta t = \delta(m \times v)$$

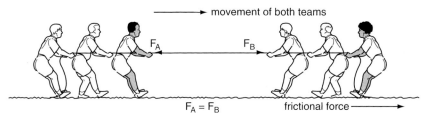

Figure 7.1

On the right hand side is the quantity (m × v) known as **momentum**. Rearranged in this way, Newton's Second Law states that force is the rate at which momentum is being increased or decreased.

- *That when two bodies interact, the forces they exert on each other are equal and opposite.* The extract is a statement of Newton's Third Law. For an object at rest this agrees with our experience. But surely the *winners* in a tug-of-war are pulling harder? If you look at Figure 7.1 you will see forces that this observation has not taken into account. The additional force is provided by the friction at the ground so tug-of-war competitors need good leg muscles as well as strong arms.

- *The forces acting on falling objects.* After dealing with the balanced situation of an object at rest the forces acting on a falling body can be considered. In the laboratory the effects of air resistance on a falling ball are negligible and pupils can see that falling objects get faster and faster. They can measure the acceleration themselves and begin to understand that all objects have the same acceleration. Its symbol is g and approximate value 10 m/s^2.

We know from above:

$$\text{force} = \text{mass} \times \text{acceleration.}$$

For a falling object the force is a measure of the Earth's gravity pulling it down i.e. its weight.

$$\text{weight} = \text{mass} \times \text{g.}$$

Think of a mass of 3 kg. Its weight will be 3 × 10 = 30 N. Think of a second mass, twice as heavy, 6 kg. Its weight will be 6 × 10 = 60 N. So a bigger force is acting on it *but* Newton's law also says bigger masses need bigger forces to move them. Double the mass requires double the force to accelerate it *so it does not accelerate any faster than the smaller mass.* All this presupposes no effects of air resistance etc.

A practical example: If acceleration depended on mass then when parachutists joined hands they would fall faster. Seeing shots of free fall shows this doesn't happen and in any case two already joined would not be able to get together with a third.

Where there are noticeable effects of friction/air resistance, the **resultant** of the weight and upward force must be found by subtracting. This explains why a piece of paper would fall more slowly than a pin of the same mass. The upward force on the paper is much greater so the **resultant** downward force is much smaller. However the *weight does not change*. On the Moon a dropped hammer and a feather land at the same time because there is no upward force (the time for both to reach the ground is more than it would be on Earth).

- *Why falling objects may reach a terminal velocity.* The situation of an object in air falling fast enough to reach terminal velocity can best be imagined in terms of parachutists. Deal with this as a strip cartoon as in Figure 7.2.

1 The weight exceeds the resistance and so he/she accelerates.

2 The resistance depends on speed so the faster the parachutist the greater the resistance until it equals the weight. There is no further acceleration, terminal velocity has been reached.

3 It is too fast to hit the ground at this speed so a parachute is opened. This increases the area so increases the resistance. The resultant force is now upwards and so the parachutist decelerates.

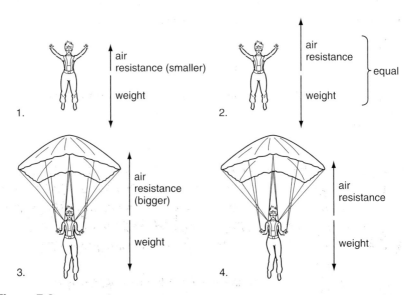

Figure 7.2

4 The resistance decreases with reducing speed until the resultant force is once more zero. The new terminal velocity will ensure a slightly safer landing!

Note: at no time does the parachutist move **upwards** or the downward force (weight) **change**. A graph is a great help in explaining this (see Activities).

KEY STAGE 4 ACTIVITIES

Balanced forces (Newton's first law)

To investigate the forces acting on a moving object without the interference of weight or friction the following are possible:

- Very low friction experiments like pucks on ice, or cushions of solid carbon dioxide, or the air track (Chapter 6). These should all show constant motion in a straight line in the absence of unbalanced forces.
- Modelling space travel where there is no friction and no gravity.
- Using computer modelling. Suggestions are given in the Resources.

Unbalanced forces (Newton's second law)

The relationship ($F = m \times a$) can be given as it is a relationship which is not at odds with our experience. If you want to demonstrate its verification then I would do it accurately using the air track. The difficulty is that of accelerating the vehicle with a *constant* force. This is achieved by pulling it forward with an elastic cord whose length you keep constant while you are pulling.

This means of course that you have to move faster as the vehicle accelerates so start with the cord stretched only slightly and have a marked point to ensure the fixed stretch. To achieve double the force use two matching cords stretched to the same marker, then three. Don't go to four unless you are particularly athletic. As explained in Chapter 6 the computer will find the acceleration for you but question pupils' understanding first. Many will believe that the constant force means constant speed, in spite of all the evidence about balanced forces you have provided. The measurements should demonstrate that the acceleration is proportional to the force applied. You can follow up with a similar experiment where the force is kept constant and the mass changed by adding to the vehicle, preferably by doubling the mass to keep the relationship simple. Double the mass needs double the force to achieve the same acceleration.

The equation can be solved by the triangle method (see Chapter 5). Provide lots of interesting examples, your textbooks will contain plenty.

A video entitled *Laws of Motion* (Granada TV) is useful here.

Momentum

This concept is useful in analysing more complex problems of forces, such as bodies colliding or objects exploding. The derivation shown in the concepts gives the momentum as the product **m** × **v**, giving it the units of kg m/s.

Simple collisions can be illustrated using the linear air track whose details are given in Chapter 6. Collisions can be modelled by allowing two of the vehicles to collide. They can collide and stick together using Velcro or a cork and pin method, representing an inelastic collision. Alternatively they can bounce apart using repelling magnets or a cone and elastic band, representing an elastic collision. The mass of a trolley need not be measured but just referred to as one trolley. The speeds can be measured using light gates. You will need to start with one trolley stationary (i.e. momentum zero) in each case, otherwise there will be difficulties in measuring all necessary speeds. The trolley that is going to move can have its mass changed to two trolleys or three trolleys to extend the range of readings.

For each situation calculate the momentum of each trolley and compare the total momentum before collision with that afterwards. This should be straightforward as they will all travel in the same direction so all values will be positive.

You should find that whether the collision is elastic or inelastic the total momentum does not change during a collision, that is, it has the same value before and after. The same result arises from explosion experiments. They can be demonstrated with the same apparatus. The vehicles can explode apart either by releasing the ones with repelling magnets or by a spring loaded device.

The principle is referred to as the Conservation of Momentum and can be illustrated by reference to snooker balls (as long as they are not subject to spin) or ice skaters colliding or pushing apart.

Action and reaction (Newton's third law)

A tug of war can be used to illustrate. Have the two pupils pull apart but remain stationary at first.

- *What forces are present?* Equal pulls in the opposite direction.
- *If one moves backwards he/she accelerates initially, so does the second. Which force does the accelerating?* Friction between shoe and floor *not* because one pulls harder
- *How can we be sure of this?* It would not be possible on a slippery surface e.g. ice.
- *Which is the unbalanced force?* The same, it causes acceleration.

But the forces of each pupil on the other remain equal. Similar experiments can be done with trolleys each with a force-meter pulling on the other. Whether stationary or moving at constant speed the forces between the trolleys are always equal to each other, that is, the meter readings are the same.

A similar situation arises when stepping off a boat. The accelerating force on the person is the friction with the boat and an equal force acts on the boat in the opposite direction. Unlike the tug of war the boat is not held in place because friction with the water is low. As the boat moves away the person falls in the water. A small clip of home video would make this point as it's one of those things people never seem to remember, especially when they are on holiday.

Since weight is a pulling force it too follows the same rules. Pupils will be incredulous to be told that the Earth is pulling on them with a force that is equal and opposite to their pull on Earth. They, personally, are too small for the effect to be noticeable to the Earth but remind them that not only does the pull of the Earth keep the Moon orbiting but the Moon also pulls the Earth and its effects can be seen in tides. Persuade them that a gravitational force exists between all bodies that have mass and that this force is what governs the positions and movements of all the stars and planets.

Falling

Start by dropping something, a stuffed toy will do, and asking what happens to the speed. Everyone should be happy to say it accelerates.

- *What force causes the acceleration?* Weight i.e. the pull of gravity
- *What is it pulling on?* The mass
- *Is it a steady acceleration?* We can answer this by slowing it down and measuring times or using an accurate measuring device for falling in air.

If the ball bearings in glycerol have not been used before try marking the tube with white tape at 10 cm intervals and getting everyone to clap each time the ball falls past a marker. The claps can be heard to get steadily closer together. This can be investigated quantitatively using stop clocks and dealing with the results graphically. If speed is plotted against time it can be shown to change by the same amount in equal time intervals.

Use the light gates again (see Chapter 6) but this time line them up vertically and drop a piece of card between them. It may be necessary to add a weight on one side to stop it twisting. Use the program as before to measure and calculate the acceleration. It can be done in a similar way with ticker timers but be prepared for quite poor values.

If objects of different mass and the same area and volume are dropped e.g. pieces of metal sheet and dense plastic, then provided none of them is too light the same value for acceleration should result. This should cause some discussion/surprise. Tell pupils that Galileo's friends were just as surprised at this result (though the story of dropping cannon balls from the Tower of Pisa is not likely to be true). Help pupils to distinguish mass (a measure of matter) and weight (gravitational pull of the Earth).

Say you are going to model Galileo's experiment with a ball bearing and a cotton wool ball of the same size. They do not hit the ground at the same time, of course. Although if you use a coin and a circle of paper and place the paper behind the coin they usually fall together as you have removed most of the air resistance from the paper. Pupils will realise that getting rid of the air will allow free fall for both items. This can be done by withdrawing the air from a long tube, using a vacuum pump. Warning – wear safety goggles and protect pupils with a screen. The cylindrical tube should be securely wrapped in tape in a criss-cross design along its full length in case any cracks have developed. Make sure the two balls or coin and paper are in the tube, evacuate it, then just invert it and the two items will reach the bottom at the same time. Traditionally this is called the guinea and feather experiment. A video of this type of experiment on the Moon is available on the Cambridge CD-ROM listed below.

Multiflash photography can be used to find a value for g, the acceleration due to gravity (see Chapter 5).

The Channel 4 programme *Scientific Eye – Gravity* has some useful images to help with understanding gravity although contains some strange ideas about space travel.

Terminal velocity

The flow chart of diagrams suggested in the Concepts above can be given in pieces to be put together in order and a suitable commentary added by pupils. An additional representation on a graph may help too (Figure 7.3).

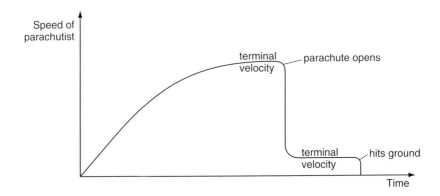

Figure 7.3

- Provide a series of drawings showing people at fairgrounds on rides and throwing things for prizes. Have at least some instances where the person or object is going upwards and one sideways. Ask for arrows showing the direction of the force of gravity in each case.
- The gravitational pull of the Moon is only one sixth of that on Earth and there is no air. Write a story about landing there, how it would feel and what special equipment you would need for survival and communication.
- Access to a computer will allow pupils to carry out a parachute investigation using a spread sheet, see *School Science Review* reference.
- Provide familiarity with Newton's laws with some straightforward questions such as those in *Physics* by Pople and Whitehead, see Resources.

ENRICHMENT AND EXTENSION ACTIVITIES

Projectiles are a good topic to investigate. A 'monkey and hunter' is a way to introduce the topic and then pupils can devise ways of projecting lightweight items at various angles, measuring times and ranges. The experiment requires an arrangement in which a circuit containing an electromagnet is broken by a pellet fired from a couple of metres away. This is done by blowing the pellet along a tube whose end is covered by a thin piece of foil which is part of the circuit. Thus the electromagnet is switched off at the same moment that the pellet leaves the tube. The electromagnet releases an object (representing the monkey) magnetically attached to it. Will the pellet get the monkey? If the pellet tube was correctly lined up to begin with, the pellet hits the monkey. The reason for this is that both object and pellet fall under gravity at the same rate even though the pellet is travelling horizontally.

Any school trips to an ice-skating rink can be taken advantage of to try out Newton's first and third laws. Take stop-watches and measure the weight in newtons of the participants beforehand. Try some explosions, pupils pushing away, with different numbers of pupils added to find some general rules. A camcorder would be a bonus here, if only to encourage participation.

More able pupils should do more graph work, and get to know the equations of motion. These can be derived from a study of the graphs while text books will provide an explanation and a variety of linked problems, see Resources.

They may also be introduced to circular motion, perhaps not quantitatively, but to understand that circular motion is caused by a centripetal force i.e. one directed along a radius towards the centre of the circle. This force can be measured in an experiment that can be found in a number of texts e.g. *GCSE Physics* by Tom Duncan.

If you are in the area a visit to Woolsthorpe Manor is worthwhile and will become more pupil-friendly if new plans for its development go through for April 2000. Try it yourself first. Details from The Custodian, Woolsthorpe Manor, 23 Newton Way, Woolsthorpe-by-Colsterworth, nr Grantham, NG33 5NR.

Crest awards on car-related topics are motivating for pupils to work on. Ford have collaborated with Crest on a car related topic called Automotivation, see Resources.

RESOURCES

Information technology
- *Motion* – a CD-ROM from Cambridge Science Media, 354 Mill Road, Cambridge CB1 3NN Tel 01223 357546.
- *Investigating Forces and Motion* – a CD-ROM from Granada Learning, 1 Broadbent Road, Watersheddings, Oldham OL1 4LB Tel 0161 627 4469.
- *Forces and Effects* by Bob Gomersall, BTL Publishing, Bradford Technology Limited, Business and Innovation Centre, Angel Way, Listerhills, Bradford BD7 1BX.

Video
- A video entitled *Laws of Motion* (Granada TV) is available from the Institute of Physics Education Department, 76 Portland Place, London W1N 3DH.
- Scientific Eye *Gravity*.

Teacher resources
- Sections 19.4 and 19.5 *GCSE Physics for You* by Keith Johnson.
- School Science Review, Dec 1997, Vol 79 (287) for spreadsheet information.
- For questions on Newton's laws see Section 1.9 to 1.13 of *Co-ordinated Sciences Physics* by Stephen Pople and Peter Whitehead.
- The derivation of the equations of motion maybe found in the first section of *Physics* by Patrick Fullick, published by Heinemann.
- *GCSE Physics* by Tom Duncan for details of measuring the centripetal force.
- Details about Crest Automotivation from Crest Awards, 1 Giltspur St, London EC1A 9DD. Tel 0171 249 3099.

chapter 8

Forces on materials

BACKGROUND

This chapter deals with the forces which act to change the shape of materials and includes the concept of pressure. Pressure, like work and force, is one of those words that pupils know from everyday experience but which needs to be carefully explained in the context of physics topics. It is usually introduced through the pressure caused by weight acting downwards on surfaces and then extended more generally to include forces at different angles and in liquids and gases. However the latter needs a rather different approach since the pressure acts outwards on the walls of the container rather than just downwards.

The action of pressure in all directions in liquids and gases will have some interesting applications, for instance, balloons, deep sea diving and hydraulics and pneumatics. The astonishing pressure caused by the atmosphere provides opportunity for popular demonstrations.

It is not one of the more difficult concepts and pupils can grasp the idea that a pressure depends not only on the force causing it but also on the area over which it acts. They will be familiar with examples varying from sinking into snow to pinning up posters. Quantitative work can be kept straightforward. The unit of pressure, the pascal, is named after Blaise Pascal, a scientist of the seventeenth century.

Pupils at Key Stage 2 will have formed a concept of stretching and compressing (elastic behaviour) from practical work with springy materials. In Key Stage 4 pupils can investigate properties of a range of materials under stress and through their load v extension graphs gain an understanding of terms such as elastic, plastic, tough, brittle.

Common misunderstandings

- pressure acts solely downwards in liquids
- following on from this, pressure is somehow linked to weight

• describing movement due to pressure differences as caused by a sucking action, rather than a pushing force, e.g. evacuating a can

KEY STAGE 3 CONCEPTS

Pupils should be taught:

• *The quantitative relationship between the force acting normally per unit area on a surface and the pressure on that surface.* The pressure that a force causes on a surface clearly depends on the size of the force applied but it also depends on whether the force is spread out over a large area or concentrated at a small one.

Mathematically,

$$pressure = \frac{force}{area}$$

Units are measured in N/m², also called a Pascal (Pa)

Example: *What is the pressure of a crate of coca-cola weighing 250 N if its area is 0.25 m²? What is the difference if two crates are stacked one on top of the other?*

$$pressure = \frac{force}{area}$$

$$= \frac{250\ N}{0.25\ m^2}$$

$$= 1000\ N/m^2$$

For stacked crates the pressure on the lower one is 1000 N/m² and on the ground is 2000 N/m².

In practice most of the areas will be more conveniently measured in cm² so ensure that pupils express this correctly as N/cm². This is another case for the triangle method of solving equations (see Chapter 5).

The inclusion of the word 'normally' in the extract is necessary to cover the possibility of forces acting at an angle to the surface in question. Since the normal force, i. e. the one at right angles to the surface would have to be found as a component using a vector triangle (Chapter 5) we need not present this to Key Stage 3 pupils.

• *Some applications of this relationship, e.g. the use of snow shoes, the effect of sharp blades.* We can divide the applications into occasions when we want to be lower the pressure and when we want to raise it. This can be reinforced through examples of design for this purpose, practical measurements of force and area and examples from 'nature'.

KEY STAGE 3 ACTIVITIES

Pressure

Begin by demonstrating with a rectangular block placed on its face with the largest area. Put a weight on it too if you want to. There is a corresponding pressure due to this weight.

- *How do I know it is?* You can feel it (or if it was on a jelly the pressure would squash it)
- *Is the pressure the same if I turn it on to this (intermediate) face?* No, it is bigger.
- *How do you know?* You can feel it – it would sink further into the jelly.
- *Why is the pressure bigger when the weight has not changed?* The area is smaller.
- *How can I increase the pressure even more?* Turn it on to its smallest face.

Work through the calculation of the pressure on each face, using a simple value for the weight.

Insist on correct setting out of calculations, and use of units. Stress the meaning, for instance a pressure of 6 N/cm² means a 6 newton force pressing down on every square centimetre.

Pupils can then work out their own pressure by weighing themselves on bathroom scales marked in newtons. They then find the area of both feet by standing on graph paper, drawing round it and then counting squares. They can then find out what the pressure is if they swing on one leg of their stool, as many of them do, given a chance.

They can try some other examples involving, possibly, elephants on one leg, drawing pins, stiletto heels, weightlifters etc. As a guide an average adult weighs around 700 N and an elephant probably 10 times as much.

Applications and examples

You will be able to find examples or pupils might suggest them. You could ask them to put them into groups, such as:

	Pressure 'raiser'	Pressure 'reducer'
Artificial	Knife	Snow shoe
Natural	Woodpecker's beak	Penguin feet

A new Science in Action programme called *Forces and Pressure* should have some more unusual demonstrations to show.

Assessing pupils' learning

- Some examples for calculation are easily made up or found to be in any physics textbook
- Explain the design of an ice-skater's boot (it would help to know that the top layer of ice melts under pressure)
- How can anyone lie on a bed of nails? Estimate the weight of the person and the area of the nails and hence the pressure. What difference does it make if the nails are filed down?
- Compare the effect of a rugby player in boots and a person in stiletto heels on a polished wooden floor – make sensible suggestions as to the weights and areas involved.

ENRICHMENT AND EXTENSION ACTIVITIES

Set a task to find a weather forecast from a newspaper and look for high and low pressure areas, linking them with the wind directions. The numbers written on the equal pressure lines are in millibars where 1 bar is one standard atmospheric pressure. This is the same as 10^5 (100 000) Pa.

Explain syphoning in terms of atmospheric pressure.

Research and report on the aneroid barometer. Ask a senior citizen if they remember having one in their house, how they used it and whether it was reliable.

KEY STAGE 4 CONCEPTS

Pupils should be taught:

- *How extension varies with applied force for a range of materials.* Stress is applied to materials by adding a load to stretch, compress or twist it. Stress is calculated from force divided by area and it produces strain, which compares the change in size of the sample to its original size. Elastic materials return to their original shape when deformed and their extension is proportional to the load applied. This relationship is called Hooke's law. It breaks down at a point called the elastic limit when the material starts to become plastic and will no longer return to its original shape. Tough materials deform but require much greater stress to do this. Brittle materials do not deform. They may be strong, like concrete, or weak, like biscuit, but they will snap rather than bend.

- *How liquids behave under pressure, including simple everyday applications of hydraulics.* In a system filled with liquid the pressure anywhere within the liquid is the same. Hydraulic machines can make use of this as a force multiplier as long as the force applied is over a smaller area than the force required – Figure 8.1 showing a hydraulic system will make this clearer.

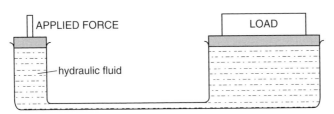

Figure 8.1

In a liquid the pressure at a point depends directly on the depth and the density of the liquid. We will probably only use water for our examples so we need not emphasise density. Pupils will understand that deep sea divers experience a greater pressure than, say, scuba divers, but they will not be clear that the pressure acts in all directions on the diver, not just downwards. She/he is not going about as if there was a 'chimney' of water on her/his head. At the risk of spraying the front row you can put a finger underneath a tap at steady pressure and any unconvinced pupil will observe or feel water coming out sideways. See activities for drier suggestions.

- *How the volume of a fixed mass of gas at constant temperature is related to pressure.* Similarly atmospheric pressure acts in all directions. It is akin but opposite to the air pressure decreasing as you go up a mountain, you don't just feel this on the top of your head. If pressure was directional in this way a balloon would shrink by flattening. This leads on to the question: what happens if we try to squash a gas?

A scientist called Sir Robert Boyle was interested in finding out what happened, so he developed an apparatus like the one we shall use which could contain a gas so that it could be squashed and its volume measured. He wanted to obtain the highest possible pressure and even used a tube of mercury the length of a mineshaft to do the squashing. He discovered the law that bears his name. It states that the volume of a fixed mass of gas is inversely proportional to its pressure provided that the temperature remains constant, meaning if you double the pressure you will halve the volume, and so on. It is applicable to the common gases at temperatures well above their boiling points.

KEY STAGE 4 ACTIVITIES

Materials under stress

You may need to begin by establishing Hooke's law if this has not been covered
earlier. Warning – pupils and anyone present should wear goggles for any
stretching experiment.

Pupils should begin with a spring that must be hung from a clamp stand with
a ruler at the side of it. Finding the **extension** will be easier if the spring is placed
with its unstretched lower end at the zero of the ruler possibly with a paper
pointer taped on to it. As weights are attached the extension can be read directly.
Reading the complete length can lead to confusion when working out the
extension. Set an upper limit to the weight that can be used, 5 N is sufficient for
most springs. The graph of extension against load in newtons will show a straight
line passing through the origin. It will work just as well if pupils use masses on
their graph but it's better to think in terms of forces.

To show the existence of a limit to the elasticity, you can stretch one spring
even further. Apart from wearing the goggles and sitting the class well back,
G-clamp the stand to the bench and place a soft landing for the weights if/when
they fall. It will be quite clear when the elastic limit has been passed because the
spring will not return to its original shape. As this requires more weight than you
might expect it probably won't be worthwhile to try and fit it onto the graph
already started, unless you are using a spreadsheet.

Pupils might then compare very thin copper wire and rubber in the form of a
thin rubber band. Careful measurements with copper may reach the plastic stage
where the deformed sample will no longer return to its original length. At this
stage it can feel strangely soft and pliable. Rubber does not obey Hooke's law but
will show a curved line graph when increasing the load and a different curved
line when removing weights. This may only be revealed if very thin rubber bands
are used. Sample results can be given for materials which are described as tough
and brittle. Aim to provide a range of extension v load graphs with applications
that exploit the profile.

Examples: *Elastic materials for bedding, clothing, sports equipment.*
Tough materials for ropes, building, bones.
Plastic materials for moulding, plasticine.
*Brittle materials such as brick and concrete that are strong when
compressed.*

Pressure in liquids

A quick demonstration of water pressure acting in directions other than down,
with the bonus of showing that depth determines pressure, is well used but

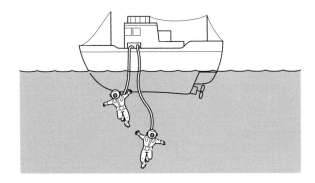

Figure 8.2

effective. Have a tall container with three to four holes drilled on one side one above the other. Station it so that the holes face a sink. Persuade a couple of volunteers to put their fingers over the holes while you fill the container to the top. On a count of three, fingers are taken from the holes and immediately you can see that the lower the hole the further the jet of water shoots out.

The principle behind hydraulic and pneumatic systems can be felt rather than seen by using two syringes joined by about 30 cm of transparent tubing. The system must be airtight. If the syringes have different diameters pupils will notice that the one with a large area is much harder to push but can make the small area piston move so far that it flies out of the end (in the air filled system). Pushing the thinner piston enables the wider one to be moved. The gain in applied force is 'paid for' by the reduced distance that the larger piston can be moved compared with the smaller. They can be used with water instead of air but this can be messy so make working over a sink compulsory.

Use diagrams to ask questions about pressure.

- Which diver (Figure 8.2) is experiencing the greatest pressure?
- Why is Ann's shower (Figure 8.3) so feeble? How could she improve it?

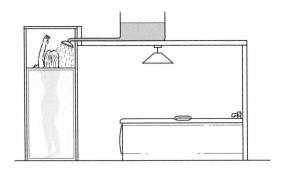

Figure 8.3

- Explain why a dam shown in cross section gets thicker towards the bottom.
- Use the equation for pressure to show that the hydraulic lift in Figure 8.1 is a force multiplier.

Left cylinder: pressure = $\dfrac{\text{force on piston 1}}{\text{area of piston 1}}$

Right cylinder: pressure = $\dfrac{\text{force on piston 2}}{\text{area of piston 2}}$

Since the pressure is the same throughout the liquid these are equal and so

$$\frac{\text{force 1}}{\text{area 1}} = \frac{\text{force 2}}{\text{area 2}}$$

$$\frac{\text{force 1}}{\text{force 2}} = \frac{\text{area 1}}{\text{area 2}}$$

So the *bigger* the *area* the *bigger* the *force* which can be achieved for a given applied force. Apply the same argument to explain how car brakes work (Figure 8.4).
- *What other simple machine is included?* A lever.
- *What liquid is used?* Oil.
- *What happens if air gets into the system?* It is more easily compressed so the brakes fail.

Hydraulic and pneumatic systems can not only multiply forces but also transfer them. This has applications in industry where danger from sparks is a problem e.g. in the manufacture of fertilisers and paint.

Pressure in gases

The apparatus for verifying Boyle's law for gases is standard in schools and should be dealt with quickly as a demonstration. It will hopefully be a fairly

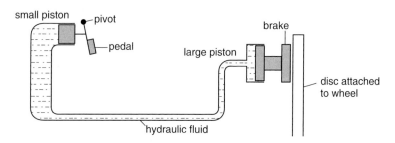

Figure 8.4

up-to-date piece of apparatus where the pressure can be read directly from a pressure gauge. The initial reading on the gauge is not zero because it is reading atmospheric pressure.

A foot pump can be attached and each time the pressure is raised the gas is squashed. As it is arranged for the gas to have a constant cross-sectional area, a reading of the length of the column of gas will be proportional to its volume. A graph of length against pressure will give a curve which is not easy for pupils to interpret. For the majority it is sufficient to understand that:

- as pressure rises volume falls
- at double the pressure the volume is halved
- at triple the pressure the volume is one third

More able pupils could plot the reciprocal of length (1/length) against pressure. You can include a 'times one hundred' on the 1/length axis to avoid plotting very small numbers. A straight line is given showing that volume is inversely proportional to pressure.

Written mathematically:

$$p_1V_1 = p_2V_2$$

where 1 refers to pressure and volume initially and 2 applies to the new values.

Lots of simple examples can be found in your school text to illustrate the use of this relationship but ensure that pupils realise that it is *only* valid for a constant mass of gas at constant temperature.

Atmospheric pressure

It's interesting for pupils to gain some idea of the magnitude of atmospheric pressure. It can be read directly from a pressure gauge but is more impressive through demonstration. If you have an unwanted ruler or thin stick try laying it across a bench with half projecting over the edge and placing several layers of newspaper over the part on the table. If you push down sharply on the projecting ruler you can break it, instead of lifting the newspaper. Amazing! But don't forget that the atmospheric pressure is over 100 000 newtons on every square meter.

As you will need to practice this you might try a more economical demonstration using a large plastic container such as a catering size container of fabric conditioner etc. Attach the container via vacuum tubing to a vacuum pump and withdraw the air. The container will crumple nicely.

- *Why does the container collapse?* Bring pupils round to the idea that it is being crushed by the atmosphere and not being sucked in.
- *What is in the container?* Nothing, so it can not 'do' anything like sucking.

Allow the air to re-enter.

- *What is the pressure inside your body?* Atmospheric pressure.
- *Why do ears pop when flying?* Because the pressure outside is less than the pressure inside and the popping and swallowing is an attempt to balance up the pressure.

Assessing pupils' learning

Investigation of cantilevers aids pupils' familiarity with patterns of materials under stress. A cantilever is a beam supported at one end e.g. clamped to a bench top. Masses can be hung on and the extension, or sag, of the unsupported end measured. Possible variables are:

- The weight in newtons (limit to 5 N for metre rules as cantilevers).
- The distance of the weight from the support.
- Length of cantilever over end of bench.
- Material or thickness of beam (it may be difficult to supply a variety here).

An interesting variation is to try applying loads to spaghetti (uncooked!). This can be done in various ways by applying the load at different points. The spaghetti can be supported vertically (in a bulldog clip) or horizontally. Be mean with the spaghetti and allow time for sweeping up. An alternative might be some thick wire.

A bridge competition is a good way to emphasise that the strength of structures depends on shapes as well as materials. Newspaper will not be regarded as a strong material but can be used, with some tape, to build a bridge capable of supporting 20 N. Specify details such as a 30 cm span, a maximum mass of a chosen value between 30 g and 50 g and no attachment to any exterior supports. It can be held up by resting at either end on blocks or stool bars. The most successful will use some sort of tubular structure. You might want to do some work in the previous lesson investigating the strength of straws to provide some ideas.

ENRICHMENT AND EXTENSION ACTIVITIES

As a follow up to the bridge competition, groups of pupils could research the styles of bridges in their area and talk about their findings to the class. Possible structures they might find include beam, cantilever, arch, suspension and cable stays.

More able pupils should become more familiar with the pascal (N/m^2) so they should be given some larger scale problems, perhaps in terms of the weights of buildings or bridges and the areas of their bases.

You may like to show the balloon in a vacuum experiment (Chapter 10) and ask for explanations. Follow up with 'Why does it take longer to boil an egg up a mountain?'

Another gas law can be introduced, that of pressure against temperature. You can verify it by demonstration where the gas is contained in a flask, which is attached to a pressure gauge. The flask is heated slowly in a water bath so that a range of readings of temperature and pressure can be obtained. The resulting graph shows a straight line but this can not be described as a proportional relationship since it does not pass through the origin. If you do trace it back to cut the temperature line then you have found the Absolute Zero of temperature (it's at approximately $-273°C$). You *can* say that the pressure is proportional to the **absolute temperature** of a gas. The absolute temperature can always be found by adding 273 to the Celsius temperature.

Example: Human body temperature is approximately 37°C or
$37 + 273 = 310\ K$

This scale is called the Kelvin scale and the ° sign is not required. Theoretically this zero point is the point at which particles have zero energy so temperatures below this, and zero itself, can not be achieved.

RESOURCES

Videos
- A Science in Action BBC programme called *Forces and Pressure* has some more unusual demonstrations to show.
- Scientific Eye *Pressure* and *Shape and Strength* from Yorkshire Television.

3

Waves

chapter 9

Behaviour of light

BACKGROUND

The study of the behaviour of light, and other types of wave, has become increasingly relevant to pupils because optical fibres, infrared remote controls, microwave ovens etc. are now everyday objects. Also studies about outer space are naturally interesting and this gives scope for considering some of the more amazing properties of light.

Ideas about how we see things, which have been introduced in Key Stage 2, will need reinforcing. Discussion of how light travels from object viewed to viewer provides an opportunity for the introduction of ray drawing. The idea of a line and arrow representing a ray of light may seem very obvious to us but it is not necessarily the case for pupils at Key Stage 3. After all we can't see light when it is travelling so how can we assume that it does travel? Perhaps it only travels in that way at night. Practice in drawing and getting the arrows going the right way should be provided in all the topics set out in this section of the National Curriculum.

Pupils will have met reflection as a process in Key Stage 2. We can build on this to investigate different surfaces (rough, smooth, black, white) and ultimately to find a rule about the angle at which light is reflected. In order to express this most usefully the 'normal' must be defined and used in ray diagrams where relevant.

An important new concept, which must be introduced here, is that of an

image and pupils should be encouraged to use it instead of the word reflection in this context.

Example: Draw your image as seen in a hall of mirrors.

These ideas, normal and image, will be used in work on refraction too. This is a process whereby light changes direction as it passes from one material to another, because it has a different speed in each material. Applications include spectacles, contact lenses, binoculars etc.

The change in direction during refraction depends on the wavelength of the light and so different wavelengths (colours) separate. We see a rainbow or a spectrum. Different colours can be investigated using filters, like those used in stage lighting. The effects caused by different coloured light is fun to study and can be quite complex if both primary and secondary colours are included.

At Key Stage 4 light should be studied much more in the context of the properties of all waves. Pupils will need evidence that it does travel as a wave. This is provided by the process of diffraction, in which light spreads out as it passes the edge of an obstacle. Under special conditions two or more beams of diffracted light may interfere in such a way as to produce darkness. Many pupils will struggle with this evidence. The major points are

1 Light transfers energy as do other waves, compare light from the Sun used in a solar cell with sound waves striking the ear drum or earthquake waves knocking down buildings.

2 Light can be reflected, compare mirrors and echoes.

3 Light can be refracted, compare the bending of light as it passes from air to water with the bending of whale calls due to different densities of water layers.

4 Detailed experiment shows that light can bend a little round corners (a process called diffraction) in the same way as sound.

5 In certain circumstances detailed below light rays can be seen to interfere, that is to produce patterns of dark and light. Such a phenomenon can not be explained if we think of light as made up of particles of any type. How can two particles arriving at the same place cause an absence of particles? However if the crest of one wave arrives at the same point as the trough of the other darkness results. This provides evidence for the wave nature of light and was generally accepted from the 1820s.

The use of a laser will demonstrate diffraction and interference very clearly.

Common misunderstandings
• light leaves the eye to encompass the object being seen

- filters change one colour into another rather than allowing a limited range of colours to pass through
- light only comes from luminous bodies

KEY STAGE 3 CONCEPTS

Pupils should be taught:

- *That, in a uniform medium, light travels in a straight line at a finite speed.* We take this for granted but do not assume that it will be obvious to pupils. We can't usually see light travelling for several reasons. We can only detect light entering our eyes and since we usually look at the object illuminated we are not at the correct angle to pick it up. If we are facing the direct or indirect source of light we think of ourselves as seeing the object or source and are not aware of light taking time to travel towards us. The reason we don't have to wait for light to hit an object is its amazing speed, $3 \times 100\ 000\ 000$ m/s or roughly 15 times around the Earth per second. Pupils like to know that we see the Sun as it was around eight minutes ago and light from the next nearest star takes four years to arrive at Earth. They will certainly want to know if some of the stars we see might have changed since the light left. The answer is definitely yes. Stars could have exploded or contracted before we could know about it but stress that this only applies to stars at very great distances.
- *How shadows are formed.* We can't see around opaque objects and light can't fall behind them. Thus shadows form. Since pupils will probably have seen shadows lengthening with position of light source, concentrate at this stage on shadows of the Moon or Earth during eclipses.
- *That light travels much faster than sound.* Light travels roughly a million times faster than sound. This means that you normally see light from a source instantaneously, however far away on Earth. Sound normally travels at about 320 m/s in air so you can detect a gap between light and sound travelling from the same distant source. Examples include thunder and lightning and cricket bat striking a ball from across a field.
- *That non-luminous objects are seen because light scattered from them enters the eye.* Non-luminous means that the object is not a source of light. To be seen the objects must be able to reflect light from another source. That light must then be able to enter the eye. Show that the source emits rays in all directions but that you are selecting just one or a few to show what happens. Pupils' own drawings will show confusion over this. They will need time to discuss and absorb the idea. The alternative view of the process of seeing as involving the eye sending out particles to somehow sense and return was commonly accepted at one time so should not be dismissed without discussion.

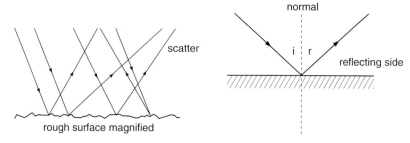

Figure 9.1

- *How light is reflected at plane surfaces.* You can use the word flat instead of plane at this stage and explain that ray diagrams show the surface as seen 'end on'. Rough surfaces do not form an image since the light is scattered in random directions. Smooth surfaces form an image because the reflection of the rays follows a simple law. The law is that the angle of incidence (marked i) is equal to the angle of reflection (marked r) in Figure 9.1. In order to mark these correctly, introduce the normal. For pupils not ready for this they can be allowed to express the equality in their own way. The normal is an imaginary line at right angles to the plane reflecting the light.
- *How light is refracted at the boundary between two different materials.* Consider light as it travels into other transparent materials from air. Although it will pass through undeflected if travelling along the normal, at other angles it will be seen to bend both as it enters and as it leaves the material. Pupils can investigate this and will find that the ray bends closer to the normal when the light is entering the material and away from the normal when leaving. This will be all that is needed for the moment. In explanation you can tell pupils that light travels a little more slowly in denser materials than in air and this has the same effect as a row of swimmers walking from shallow to deep water. Those in deep water move more slowly than those in shallow water (Figure 9.2). The diagram explains how this makes a row of swimmers change direction. An alternative model can be given using ripple tanks (see Chapter 11). But beware – ripples travel more slowly in shallow water!
- *That white light can be dispersed to give a range of colours.* White light is a range or spectrum of all colours and dispersion describes how the colours are split up by refraction. The light bends when entering a dense material but, because of the angle of the next face, is not allowed to recombine as it does when it leaves a rectangular block. This is achieved by using a prism (see any physics text). Note that blue is the colour (shortest wavelength) which is turned through the largest angle and red through the smallest. You can explain that

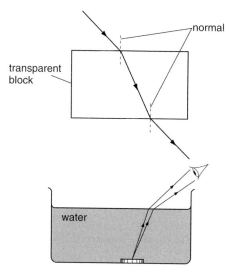

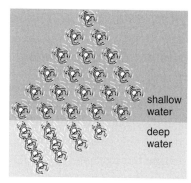

Figure 9.2

raindrops can act like tiny prisms but, like placing the screen in the right place to catch the spectrum, you have to be standing at the correct angle to the Sun's rays to see the rainbow.

- *The effect of colour filters on white light.* Through discussion ensure that pupils understand that the filters filter out some colours and allow the one(s) you see to come through. Some will believe that the filter is able to change one colour into another. Use them in conjunction with a spectrum to show their effect.

- *How coloured objects appear in white light and in other colours of light.* A red object is said to be red because it reflects red light to your eye. The same applies to green and blue objects because these are the primary colours (though not the same as those taught in art lessons). A black object reflects no colours particularly if it is matt. A white object reflects all the colours that shine on it. Three secondary colours are specified, magenta (reflects blue and red), cyan (reflects green and blue) and yellow (reflects green and red). Avoid talking about mixing colours as it becomes confused with paints. Use the example of stage lighting instead. Encourage all this type of question to be dealt with by drawing. All three colours mixed together give white, though in practice it may look off-white.

KEY STAGE 3 ACTIVITIES

Straight line propagation

Ensure that pupils have an understanding of transparent and opaque. Review their earlier work and ensure that they are confident with the idea of light travelling outwards from a source in all directions. At this stage consider only air/space as a medium and use torches in darkened rooms, sunlight through holes in the blinds etc. to show light travelling in a straight line. Really good blackout with a bright parallel beam is needed. The beam we can not normally see is shown up by smoke from a smouldering rope or talc dust.

Ask pupils in groups to do a poster sized drawing to show how they see a lamp. Through discussion and agreement convert the drawing into a ray diagram. Always include the pupil's head in the diagram, later it can limited to just an eye. Follow up with a torch in a darkened room. A beam is a bundle of rays. Point at a pupils face but not with a bright torch!

- *How do we know where the beam goes?* We can see where it arrives. It reflects off the face I am pointing it at.
- *Why can't we see the rays as they travel through the air?* They don't reflect off anything. But the owner of the face can see them.
- *What if we put something in the way?* Use smoke from a smouldering straw or talc powder to pick out the beam. It is seen to be straight.
- *Do the rays ever bend without hitting something?* No.

Pinhole camera

The pinhole camera demonstrates straight line propagation of light in a simple and effective way. Two stages in its investigation are:

- the formation of an image
- the storing of the image on light sensitive material

Introduce the idea of an 'image' as what we actually see and insist on its use in future. Demonstrate how to make the camera and allow pupils to investigate the images as follows. You will need a darkened room though not too dark as bright light sources can be used at this stage. A bright bulb with a large clear filament is best, candles and bunsens are also possibilities but, apart from the hazard, the flames tend to move about with air currents. If you are lucky enough to have a sunny day you can look at an image of the window. Figure 9.3 shows the arrangement and also how the rays are drawn to explain the appearance of the

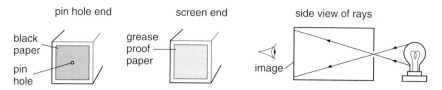

Figure 9.3

image. Note: check that students view from the screen end, it's usual to peer through the hole instead.

1 View the image and the effect of moving towards and away from the object. There will be surprise that the image is upside-down but hopefully the ray diagram will clarify the situation.

2 Make three or four more pinholes – three or four more images will be seen.

3 Make one large hole, combining the pinholes – the image will be very blurred and pupils will realise that a pinhole is needed for a clear image.

4 Place a small convex lens in front of the spoilt hole and the sharp image will be magically recovered. Use this to raise the idea of lenses altering images by bending light rays.

5 Is it in colour? Again this will be quite a surprise, but do forestall disappointment if you take photographs as you will almost certainly use black and white film.

Storing the image is really only possible if you have access to a good sized darkroom or can arrange for small groups of pupils to work on this while the remainder do research work on, say, cameras. It can work extremely well provided the pinhole cameras are light-tight with small, sharp pinholes. In red light pupils replace the screen with a piece of photographic film, close the box and cover the hole with a thumb or finger. If a colleague carries a stool outside the camera can be placed on this and then the hole uncovered for a fixed time. On a sunny day with the back to the Sun about 25 seconds is sufficient but will depend on the film used. The hole is re-covered and taken back to the darkroom. The sequence for processing is developer to fixer to rinse in water. The time in the developer depends on the brand that is used but as the picture can be seen as it appears, it can be rescued before it gets too dark. Good pictures can be taken but it is worth encouraging those with foggy results that it is probably light entering so not really their lack of skill.

Shadows

You can demonstrate an eclipse of the Moon quite simply using two balls, one whose diameter is about one quarter of the other. You need a rather extended light source for the Sun, an overhead projector is good. Explain that you can not show distances to scale with this model and also that the Moon does not orbit the Earth in the same plane as the Earth's orbit around the Sun so we don't see an eclipse in every orbit. Due to the tilt of the Earth total eclipses of the Sun happen rarely in the UK. The next after August 1999 will be in 2090.

Speed of light

Discuss the phenomenal speed of light (300 000 000 m/s) and what it means in terms of seeing stars. It is not difficult to calculate the time taken by light to arrive on Earth from the Sun if pupils are confident using calculators for large numbers. Make it slightly easier by working in km.

$$\text{time for light to travel from the Sun} = \frac{\text{distance from Earth to Sun}}{\text{speed of light}}$$

$$= \frac{150\ 000\ 000}{300\ 000}$$

$$= 500 \text{ s}$$

Some pupils will want to work out for themselves how far in km a light year is (300 000 × 60 × 60 × 24 × 7 × 52). Make sure that any references to a light year are consistent with it being a *distance* unit not a *time* unit.

The enormous difference in time taken by light and sound to travel terrestrial distances allows the measurement of the speed of sound. This is described in Chapter 10.

Scatter and reflection

Ask pupils next to draw a ray diagram on a poster for seeing a non-luminous object, say, a book. Again, by discussion and consensus, arrive at a suitable ray diagram. Include rays which do not go to the eye to show scatter. This would look better on an OHP where the scattered rays could be on an overlay. You could then add and remove them to avoid confusion.

- *Some surfaces reflect light better than others. Which ones?* Light, shiny ones, mirrors etc.
- *What do we really see when we look into a mirror?* Work towards 'an image of ourselves'.
- *How does that image get there?* Because of the light.

- *How can we investigate what happens to light when it hits the mirror?* Arrange a narrow beam of light to hit the mirror and see where it goes next.

Find the position of an image first. You can demonstrate using a well-supported piece of plate glass and two candles. If pupils do it, it is safer to use white chalk crosses on black paper in subdued light. Try to place the second candle/cross in the same place as the image of the first that you can see looking into the glass plate. Three points should be emphasised.

1 Introduce the idea of an **unreal** (virtual) image. You can see the image of the candle by looking into the glass but you could not put it on a screen. This is different from the pinhole camera but the same as a mirror.

2 The image is the same size and way up as the object candle or cross.

3 The image is the same distance behind the reflecting surface as the object is in front.

Challenge pupils to find a rule about the way light is reflected from a surface. Your school will have suitable ray boxes and it will be worth spending a bit of time beforehand so that you can advise pupils on how to get a nice bright and long beam. Before using the ray box get pupils to set the mirror upright on a piece of white paper and draw a line to mark its position. Pupils can then draw a line at right angles to this to be the normal. They then point the light at the meeting point of normal and mirror. It can be seen that the light seems to reflect off at the same angle at which it arrives, whatever the angle of incidence. Depending on ability the directions can simply be appreciated or marked and carefully measured and compared. Any pupil can claim the resulting law as their own but will have to accept that they are not the first to discover it.

Finish off with Pepper's ghost. The arrangement is similar to that above but place a beaker of water behind the glass. The result shows the image of a burning candle apparently under water. Tell pupils this was used on stages in the last century to reflect the image of a ghost onto the stage. Though very effective, actors were not keen as they had a tendency to walk into the large plate set diagonally across the stage.

Mirror writing is fun to try. Pupils should be able to explain why it appears on ambulances and some police cars.

Refraction

Rectangular blocks of glass or Perspex can be used to demonstrate refraction. Holding them up to see how they distort the view, and other peoples eyes, leads into the idea of images made by bending light i.e. it seems to come from a different place from that expected. Water will perform similar tricks. Rulers look bent and coins can be made to appear. Have everyone sit around a pot (not transparent) which has a coin in the base. Tell pupils to sit so that they *just* can't see the coin. Pour in water, gently so the coin doesn't move, and soon everyone will be able to see it without moving. Ask the fishing fanatics in your class about apparent depth of fish. A ray diagram is shown in Figure 9.2.

Dispersion

Pupils can get rainbow edges around their friends by using Perspex prisms. To get a really good spectrum a good quality prism is needed. Use the relatively narrow beam from a slide projector and set the prism on its triangular base. Turn it and ask everyone to look for the spectrum (rainbow). In a somewhat darkened room it will show up well on someone's shirt. Place a screen in this direction. Examine the colours and try to decide how many there are (Newton chose seven). Stay with this arrangement for the next section on colour.

Colour filters

Use coloured gels to show the effect of filtering out parts of the spectrum. If stocks allow let this be a class investigation using ray boxes. Note that pure red, green and blue will transmit only one colour (expect the blue to be rather faint), while the secondary colours transmit more than one as the next section shows. Show what happens when a red and green are used together – no light can get through both. Some pupils may want to know what happens to the light that is not transmitted. It is absorbed by the filter, which increases its internal energy as a result. The energy will be transferred to the air from the filter.

Coloured objects

Good sized lights should be used here for preference, such as car headlights or projectors each with a primary colour filter in front. An easy activity to begin with is to write a message on a white board in proper board writer (surprisingly good primary colours). If written in three colours it can be made to give different messages depending on the colour of the light shining on it. For instance the red part will disappear when a red light shines on it because both board and message appear red. Words in blue and green appear black in red light because they don't reflect the colour falling on them. There's a good video in the *Scientific Eye* series showing some bungling bank robbers who have not appreciated these points.

A lot of new words, or words used in a more specific way, have been covered in this topic so an illustrated vocabulary on the wall would be a useful aid.

Preparing one would be a good way of revising the section. Another promising resource is a video from the BBC's Science in Action called *Light*.

Assessment/assignments

Drawing and colouring objects as they would appear in, say, red light e.g. flags, party decorations.

Pupils can construct ray diagrams of optical systems such as periscopes and kaleidoscopes showing how they agree with the law of reflection, or to simply describe how they used a periscope to see their favourite group at a concert.

What types of images are found in spoons? – three distinct types will be found by careful observation but if the spoon is particularly concave the magnified image will only appear if you hold it very close to your eye.

Where must you dangle a magnet to pick up some lost keys? Provide a diagram of the keys at the bottom of a pond for pupils to complete with rays showing where they seem to be.

ENRICHMENT AND EXTENSION ACTIVITIES

Investigate the glasses worn by friends and family – what types of images are found in the lenses worn by short-sighted/long-sighted people?

Many pupils will not find ray diagrams intuitive and they should only be given the simplest ones. They can spend more time looking at what plane mirrors can do, designing where to put a mirror so that you can watch who comes into a door for instance and working out image distances as objects move closer or away.

For those who can accept the model they can spend more time looking at ways of bending light using different materials such as lenses. Some schools may have Perspex lenses, which lie flat on paper. Used with a ray box, they show how rays are bent as they pass through a lens because their lower surface is coated white to show them up.

The idea of a focus is useful and can be shown on a sunny day by focussing rays on to paper. To get a focal length the rays need to be from a very distant object, i.e. the Sun and then the distance is measured from the lens or mirror to the focal point.

KEY STAGE 4 CONCEPTS

Pupils should be taught:

- *That light can be reflected, refracted and diffracted.* Extend the Key Stage 3 study of reflection to examining images in combinations of plane mirrors and a look at curved mirrors. According to the optical arrangement in use, images can be magnified, reduced or the same size, they can be upside down or upright and they can be real or virtual. Real images can be put on a screen because the light *really* arrives there. Virtual images cannot because the light only *appears* to come from there, i.e. behind a mirror.

There is a law for refraction dependent on the materials either side of the boundary but it is not necessary for pupils at this stage. They should be able to do careful drawings showing the rays of light at water/air or glass/air boundary, such as that in Figure 9.2.

Diffraction will be a new phenomenon for most pupils and is more easily understood if it follows demonstrations with the ripple tank. The term is applied to light, and other waves, spreading out as they pass around an obstacle or through a gap. The gap or obstacle forms a starting point from which a new wave sets out.

Travelling as a wave raises new possibilities, that of different parts of the wave arriving at the same time at a point. If say, a crest on one wave arrives with a trough on another these might be able to 'cancel out' to give darkness. In practice the two waves need to set out from the same source and travel paths which differ by half of one wavelength. If the **path difference** is zero or one complete wavelength the waves add to give a bright spot. The effect of waves adding or subtracting at the screen is called **interference**. The resulting pattern for light waves, usually formed from slit shaped gaps, is a pattern of lines, known as fringes.

As with refraction the amount of spreading out depends on wavelength so this, too, produces a spectrum. Since the wavelength of light is very small the effects are less familiar than reflection and refraction. Look for diagrams to illustrate diffraction in Chapter 11.

- *The conditions for total internal reflection and its use in optical fibres.* Refraction studies showed that a ray of light leaving a denser material to move into air bends away from the normal, but what happens when the angle of refraction reaches 90°? At this stage the angle of incidence is approximately 42° and is called the critical angle for glass. If you try to increase the angle of incidence you will see the surprising result that the light is now completely reflected within the glass allowing none to escape. The effectiveness of this reflection is the basis for the transmission of light along optical fibres.

- *Some uses of visible light in communications.* Apart from optical fibres pupils will be able to suggest plenty of examples here, both low-tech and high-tech, from bonfires to lasers. In order to use the highly efficient optical fibres for communication, analogue electrical signals from, say, a microphone are converted to digital signals and transmitted as pulses of light along the cable. Analogue signals are ones which can have *any* value, digital ones can only have two values which are obtained by frequently sampling the analogue signal. The pulses are received by a light sensor which transmits them to a device which reconstructs them as an analogue signal to drive a loudspeaker. Such fibres are cheap, lightweight, efficient and capable of carrying many times more information relative to copper wires. They can also be made extremely narrow and so are used in endoscopy, reducing the need for biopsy or surgery.

KEY STAGE 4 ACTIVITIES

Reflection

You can see images in curved mirrors by simply looking into the back (convex) or front (concave) of a spoon. Distorted and inverted images can be examined more carefully setting up an object (optical pin) in front of a curved mirror and describing the image that results. The image in a convex mirror (driving mirror, shop security mirror) is always smaller than the object, the right way up and behind the mirror i.e. virtual. The image in a concave mirror (cosmetic, dental) depends on the distance of the object. If it is closer than the focal point the image is magnified, virtual and upright, if further away, the image is smaller, upside down and real. More able pupils might be interested to see the ray diagrams that form these images but it is rarely called for in GCSE syllabuses.

Refraction

Similar activities can be carried out in the study of lenses. Pupils find the formation of an image of a window and the view outside on to a page of their book surprising. Convex lenses should be used for this and it's most effective on a bright day. You can measure the distance from the lens to the point of clearest focus to give the focal length. Concave lenses only form virtual images. Studying print or graph paper through each lens, and through some with different focal lengths, will illustrate the type of image formed.

The most important use of the properties of lenses is in spectacles or, even earlier, in the eye itself. In studying the eye and its corrections we can model what is happening to the rays instead of just looking at the image. The model of the eyeball is made from a large round-bottomed flask supported on a ring and filled with a flourescein solution. The solution reflects the light so that we can follow the beam, provided the room is pretty dark. Three lenses should be attached to the outside of the flask, each will bring light to focus at the apex of cone. Provide light from a slide projector. Choose one lens, representing normal sight, to bring the light to focus just on the far side of the flask (the retina). A second lens, showing short sight, shows the focus is short of the retina. A suitable concave lens will correct the focus on to the retina. Use the other lens to make an image that falls outside the flask. This long sight can be corrected with a convex lens. If you have a sight defect you can use your own spectacles, or some discarded ones, to check their correcting effect.

Diffraction

In your scheme of work arrange for demonstration of diffraction using the ripple tank to precede its demonstration with light. It's much easier to understand that a wave motion sets out anew at an obstacle. See Chapter 11. You can also bring out the dependence of the effect on the relative sizes of obstacle and wavelength. A gap needs to be of a similar size – if too big it represents no obstacle to the much smaller wavelength; if much smaller, insufficient energy passes through to allow the detection of a diffraction pattern.

Diffraction can be detected in light only for gaps of the order of 10^{-7} m so is not an everyday experience. Pupils can see it simply by holding close, but not touching, their thumb and first finger. View the gap against a bright window. Bring them slowly together. Just before they touch a bump appears between the two, caused by the diffraction of the light at the narrow gap. Looking through an open weave fabric at a small light source in a darkened room provides some interesting diffraction patterns.

Observing a diffraction pattern using white light and two slits is best left to Year 12/13 pupils but two ways which are appropriate are:

1 Using a diffraction grating – this is a whole row of fine lines on a transparent sheet. Light passes through between the lines and then interferes, forming not only bands of dark and light, but a spectrum because of the dependence on wavelength.

S 2 Using a laser – popular but for teacher's operation only. Most school lasers are very low powered and do not constitute a hazard. However safety routines must be adhered to, maintaining good practice. Check out the manual provided, its main point will be that pupils must not look directly at the beam or its reflection from a shiny surface. They will already be aware of its potential from its use in science fiction so will be prepared to show a healthy respect. To this end sit pupils behind the laser and place a translucent screen for the reflection. Small parts of the beam may be scattered so that they reflect from walls, etc. but this is not a cause for concern. The presence of the beam of red light travelling from laser to screen can be shown by sprinkling talc or spraying water in its path. Show diffraction by setting in the path of the beam a single slit. Pupils will notice that the result on the screen shows not only spreading out but also banding caused by interference between light from different points on the wave front. You can also show patterns arising from two slits, a grating and even reflection on to the screen from a CD, a reflection grating. Take care that the reflection is upwards onto the ceiling, not towards pupils. The point to bring out is that between the bright spots there are dark patches. From the explanation in the Concepts you can use this to bring out the wave nature of light.

Total internal reflection

The equipment used here is generally the semi-circular glass block or prism so that measurements of the critical angle for glass or Perspex can be made. This is done using the ray box. The results of total internal reflection can be demonstrated using any longish piece of transparent material, such as a glass rod and a bright light source. Don't expect to see the rays within the material only emerging from the ends. Some long pieces of optical fibre will show that transmission is equally effective when the cable is wound up, twisted, knotted etc. Pupils may bring torches and toys using optical fibres. Coloured light is particularly effective.

Assessing pupils' learning

- Design a cooker for use in a sunny climate where fuel is very expensive, a simple concave dish with the food at the focus.
- Think of/draw some examples of concave shapes being used to focus other types of waves, you might discuss helpful ideas such as ears, satellite dishes, radio telescopes.
- Draw a diagram of a fish's view of the world outside the pond and extend the principle to a 'fish eye lens'.

ENRICHMENT AND EXTENSION ACTIVITIES

Pupils can set up combinations of lenses to make models of microscopes and telescopes. Full details are supplied in the resource given below.

Interference can be investigated in soap bubbles and soap films, so that more able pupils can explain the effects in terms of what they already know about reflection, refraction and path difference.

Research into modern cameras and compare them with the pinhole camera.

Research Polaroid and liquid crystal display. This highlights a new property of waves called polarisation. In this effect transverse waves are limited to vibrations in one plane and it is caused, in the case of light, by passing it through a narrow 'grid' of aligned molecules.

RESOURCES

Video
- BBC's Science in Action programme called *Light*.
- Scientific Eye *Seeing the Light* and *Seeing and Believing* from Yorkshire Television.

Teacher resource
- Microscope and Telescope in section 11, *GCSE Physics*, Tom Duncan, published by John Murray.

chapter 10

Sound

BACKGROUND

Pupils normally enjoy this topic. The properties of sound are interesting and the investigations present opportunities for making noise legitimately. It is not conceptually difficult and contains a very small amount of numerical work. The language and terms used are in everyday use and the phenomena investigated are related to a pupil's everyday experiences. Finally, sound and ultrasound have many relevant applications, some of which are described below.

At Key Stage 2 pupils will have gained experience in the ways in which sound is made and realise that a vibrating object is a source of sound. They won't mind a repeat of some demonstrations they may have seen already, together with some new ones. At Key Stage 3 we extend this understanding into the transfer and reception of sound.

Reception of sound by the eardrum and by the diaphragm of a microphone can be examined and compared. The sound received by the microphone can be 'seen' by feeding the signal into a CRO. The idea of 'viewing' the shape of a sound is very important in establishing the nature of sound as a wave and also shows how louder sounds have taller waves, higher sounds have more waves in a given distance.

Investigations of pupils' hearing ranges show the presence of sound waves of higher frequency than the human ear can detect. Very high frequency sound is called ultrasound. It has become useful in medicine as a diagnostic tool that is considered to be less dangerous than X-rays and a treatment for some disorders such as kidney stones. It has uses varying from range finding and echo-sounding techniques to mechanical cleaning of delicate instruments and jewellery.

Common misunderstandings

- No need for a mechanism to explain how you can hear a sound from a distance – you can hear it because it makes a noise!
- Confusion between pitch and loudness.

KEY STAGE 3 CONCEPTS

Pupils should be taught:

- *That sound waves cause the ear drum to vibrate and that different people have different audible ranges.* The range of hearing for humans is roughly 20 cycles per second to 20 000 cycles per second but falls markedly with age from early teens.
- *The effects of loud sounds on the ear.* It is important that you stress this for pupils. They should be aware that louder sounds (measured in decibels, dB) cause more damage and that the damage also depends on the length of time of exposure. The damage leads to deafness later in life. The effects of personal stereos and the loudness of sounds in cinemas and discos should be discussed giving pupils an opportunity to say how they would provide advice to others. Providing the advice yourself will almost certainly have the opposite effect to that desired.
- *That sound waves cannot travel through a vacuum.* Since hearing a sound involves a wave being transmitted by moving particles, it follows that absence of particles means that sound can not travel. Hence sound does not travel above the Moon's surface.
- *The link between the loudness of a sound and the amplitude of the vibration causing it.*
- *The link between the pitch of a sound and the frequency of the vibration causing it.* A sound wave is carried by particles in a medium moving to and fro. The amplitude of a wave is the maximum distance moved by a particle from the centre of the movement. The louder a sound is, the greater this amplitude is. The more quickly the particle oscillates the more complete cycles can be fitted into one second. This is called the frequency, measured in cycles/second or hertz (Hz). The higher the frequency of an audible note, the higher its pitch or position in a musical scale.

KEY STAGE 3 ACTIVITIES

Vibrations

From a model of the ear pupils can see the similarity between the eardrum and a musical drum or tambourine, appreciating that the function of the diaphragm is to vibrate.

You might begin by asking *how can you hear what I am saying?* The ensuing discussion, preferably in groups first, should enable you to consider vocal cords, which you can feel vibrating, and eardrums and the need for a way for the vibrations to get from one to the other.

If you feel that pupils would benefit from enhancing their ideas about vibrating to make sounds you can investigate:

1 Tuning forks (particularly convincing if touched to the earlobe or just dipped onto the surface of some water). *Only* tap them on rubber bungs.

2 Loudspeakers (connected to a signal generator, the vibrations at low frequencies can be seen, at higher frequencies place some polystyrene balls in the upturned speaker cone.)

3 Any musical instrument.

Have available the biggest model possible of an ear. The drum will probably be missing so check beforehand or have ready some tissue or balloon skin as a replacement. A drum/tambourine is handy for comparison. Bring out the following points while looking at the model.

• The external part of the ear is not sound sensitive. It helps to focus sound and to identify direction.
• Formation of wax is a natural part of the ear's cleaning mechanism. It's better than poking things in your ear.
• The drum vibrates exactly in response to the sound arriving at it – point out its fragility.
• The tiny bones (smallest in your body) vibrate in the same way passing vibrations to the inner ear.
• The canals of the cochlea (meaning snail) contain liquid which help to maintain balance and a sense of position. The tiny hairs on the walls of the canals detect vibrations and are able to convert the pattern to tiny signals that pass to the brain.

A broken or disused microphone or telephone earpiece will also have the equivalent of an eardrum. It would be useful to see this before connecting a microphone to the CRO. The electromagnetism section of a standard text will contain details of the microphone.

Range

Ideally the setting up of the CRO can be managed beforehand and then just switched on when you are ready to use it with the microphone (see Chapter 2). The sensitivity should be such that you can speak into the microphone and get a clearly visible trace, you will probably need the highest setting for this. It's good for pupils to see how the sound you make becomes a picture on the screen. Sing a few notes and they will see the effect on the shape that we can call a wave. Interview a pupil, though you will find that they become strangely quiet when asked to speak. It is possible that the appearance of transverse waves on screen will be confusing if we have just learnt that sound is a longitudinal wave so explain that it is the microphone which changes the signal as it converts from sound to electrical energy. Loudspeakers do the reverse.

A machine that provides a range of frequencies is called a signal generator. Connect two leads from its output to a loudspeaker and just check that the note is at a reasonable level of volume and pitch. This may be controlled by a volume knob or possibly a power or volts output and a frequency dial. Don't maintain a single note for long, it will probably cause headaches. Two more leads may be taken from the output, this time to the input of the CRO so that the note you are listening to may be 'seen'. This will be important as you raise the frequency of the note to the point where it can no longer be heard. Do this gradually and ask pupils to raise their hand when they can no longer hear the note. Warn them that you might switch off the generator to check that they are not cheating. Most pupils will achieve 18 000 c/s. Most teachers won't. When last measured mine was a mere 12 000 c/s, much to pupils' delight. They decided that if they talked at very high frequency I would no longer be able to hear them. The appearance on the screen indicates that a signal is still being given out and you can complete the investigation by decreasing the frequency gradually to find out whether the same people pick it up at the same pitch as before.

Ear damage

This is best dealt with while the model of the ear is available, where it can easily be seen that the drum and the three adjacent bones are vulnerable to mechanical shock. Parts of the ear are also readily infected particularly in the young and this, too, can lead to deafness. Other causes of deafness include damage to the auditory nerve, defective hairs in the cochlea and welded middle ear bones. Trying to imagine deafness is difficult because you can not just put on a blindfold as in simulating blindness. Pupils should try to imagine the isolating aspect of deafness, perhaps through writing a week's diary supposing they experienced temporary deafness or through the experience of someone they know.

Transmission of sound

Ask pupils how the sound travels from the source to their ear. They will have some ideas about the pathway between and these can be confirmed/corrected using the Slinky spring. If you have a nice long spring demonstrate along the floor, if not a long bench will do. On two or three coils towards the middle attach some brightly coloured tape. Explain that one end represents the source so it vibrates. These vibrations push the coils to and fro. The far end is the receiver, the coils are the air particles in the space between. Repeat the push at regular intervals so that a wave is set up. You will see it being reflected at the far end. The coloured tape shows the movement of individual coils i.e. particles. Describe how the particles in materials are held together by forces which act like springs, so in this way sound can be passed on. This is a **longitudinal wave**.

- *Do you think sound travels better in solids and liquids than in air?* Yes, the particles are more strongly held together – hence native Americans listened to rumbling hooves through the ground. You can verify this using a ticking watch and an ear pressed to the table in a very quiet room. Supervise closely so no-one is tempted to bang the table, a painful experience for the listener.
- *Does sound travel through a vacuum, empty space?* The concept of a vacuum may be a new one to some pupils so it's worth doing one or two other experiments to show that this is not a piece of equipment just produced for this 'trick'. Before you begin, ensure that pupils are well protected from any evacuated container by a clear screen and the containers to be evacuated have strips of clear tape on in case there are any tiny fractures which could lead to explosions (strictly implosions but just as dangerous). You should wear safety goggles. Make sure you are familiar with which tap to use to set up and release the vacuum. Try the following:
- Put a balloon with a little air in, and tied, into the container, a bell-jar. Turn on the vacuum pump and as the air pressure goes down the balloon will blow itself up.
- Put a small beaker with about 50 ml of warm water into the bell-jar, having let someone dip their fingers in to check the temperature. In the vacuum the water will come to the boil. Take it out after allowing air back in and dip your fingers in. Everyone will be surprised that you are not scalded but the temperature has not changed.

A careful explanation of these is not really required at this stage although you can give one to pupils who are comfortable with ideas about particles. All that is required is an understanding of a vacuum as nothing at all.

Follow this with a demonstration that sound does not travel through the vacuum. Most schools will have one set up and kept for this purpose. The source

127

is an electric bell, which should be hung on wires to conduct and provide as flimsy a support as possible. The jar should have Vaseline smeared on its base. Turn on the bell, it can be clearly heard. Pump out the air and the ringing of the bell will fade. Pupils can see that the clapper of the bell is still behaving the same way as before. The noise of the pump tends to obscure the effect but when it is switched off and really quiet, gradually allowing the air in is very effective. Some pupils will object that they can still hear the bell even after the jar is evacuated, encourage them to supply their ideas of why (incomplete evacuation, transmission through the wires and base).

Words associated with waves

Setting up the CRO, signal generator and loudspeaker as above you can investigate with pupils the way in which loudness and amplitude are linked, similarly with pitch and frequency. These words should be used consistently now whenever talking about waves. Amplitude can also be linked with energy. A higher amplitude, louder wave will appear taller. A higher frequency, higher pitch wave will show more waves per cm of screen.

A video is useful here as they can easily show the movement that is essential to understanding waves, see Resources.

Assessing pupils' learning

- Supply a drawing of the ear and ask pupils to colour the parts that vibrate when hearing a sound.
- Pupils should use what they know about pitch, size and frequency to discuss why some animals have a higher hearing range than humans.
- Pupils should think about communicating on the Moon where there is no atmosphere and describe several possible ways. Touching helmet to helmet or via a rod is not commonly thought of but demonstrates that sound travels through materials other than air.
- Investigate musical instruments – how do you change its pitch and loudness? Design one of your own explaining how it is played. A collection of pictures of unusual instruments is a useful prop here.
- An investigation into the length of a tube and the pitch of the note it can produce can be carried out. Recorder players will know what the notes are and their frequencies can be found on tuning forks.

ENRICHMENT AND EXTENSION ACTIVITIES

Use ideas about the speed of sound to do calculations, e.g. for the distance of thunderstorms. Assuming that the arrival of the light is instantaneous (a point for discussion), sound travels at around 1/3 km per second and so three seconds between the lightning and thunder means the storm is 1 kilometre away.

Design and carry out *quiet* hearing tests on partners. An example could be dropping a pin onto a tin from same height each time to replicate the sound. You may have to negotiate some quiet spaces in the corridor for a while. Blindfold the partner and find out at what angle hearing is at its best/worst. This investigation taxes the ability to think about setting up a fair test and whether repeats will improve results.

Investigate sound travelling in solids by making string telephones, listening through wooden rods, etc. and in water by filling a large sink and listening to a tapping noise with a stethoscope. The speed of sound is higher in liquids, and highest in solids. Longitudinal waves are transmitted more quickly the more strongly bonded the particles.

KEY STAGE 4 CONCEPTS

Pupils should be taught:

- *That sound can be reflected, refracted and diffracted.* Review pupils' ideas about reflecting sound. Try to ensure that they use the word echo to mean this and not in the sense of a copy. Ask for an explanation of echoes and link up their knowledge about vibrations with avalanches. Once pupils understand that sound takes time to travel from one place to another they will be able to tackle questions about echoes. They need reminding that to hear an echo sound has to travel there *and* back.

  Since sound travels at different speeds in different materials, refraction does take place but cannot be successfully demonstrated in the school laboratory.

  Diffraction is a much more familiar occurrence with sound than light (see Chapter 9). Typical wavelengths in the audible range are 1–2 metres so diffraction is easily detected through doorways, windows etc. Interference patterns too can be detected by collecting sound waves from two loudspeakers. Regions of loud and soft can be detected.

  Relate the characteristic properties of sound as a wave (see Chapter 11).
- *About longitudinal and transverse waves in ropes, springs and water.* The difference between these two types of waves is in the direction of oscillation of whatever particles are vibrating. For longitudinal, the direction is along the direction of movement of the wave, for transverse it is at right angles.

Longitudinal examples
- sound
- worms (not exactly but they give the right message)

- p-waves (those which travel fastest in earthquakes)
- slinky (used so that the coils are pushed forward and pulled back)

Transverse examples
- ripples in water
- light and any other member of the electromagnetic spectrum
- sidewinder snakes
- Mexican waves
- slinky used so that the whole spring is swayed side to side

- *About sound and ultrasound waves, and some medical and other uses of ultrasound.* Details of these uses are to be found in school texts, or in any case, form the basis of a research and report back with diagrams for groups of pupils. Include the following:

  1 Sonar devices, where sound waves are emitted and collected after reflection, provide a way of measuring distances, assuming the speed of sound in the medium is known. The technique can be used for mapping the sea floor, detecting some geological deposits or shoals of fish etc.

  2 Using ultrasound in a similar way for mapping internal parts of the body. The echoes are mainly returned from the interface of bone and soft tissues. This is used in diagnosis of possible foetal problems, growths on bones etc. Higher energy waves can be used to actually break up items such as gall stones providing a quick and almost painless remedy.

  3 Similar frequencies can be used to dislodge dirt from jewellery and delicate machinery.

KEY STAGE 4 ACTIVITIES

Speed of sound

Dependent on the geography of your school one of two ways of measuring the speed of sound will be suitable.

- *Using echoes.* This is suitable if your school has a large area of wall at a distance of at least 100 m to cause echoes but no adjacent walls to cause confusion. You will hear a clear echo of a handclap and should then try to adjust the clapping so that the next clap comes at the same time as the echo of the previous clap. Lots of students can help with timing a number of claps, say ten, to find the time taken to travel to the wall and back. Obviously only one person can clap. Two reliable students can be given the task of measuring the distance. The speed can be calculated from

$$\text{speed} = \frac{\text{distance of wall} \times 2}{\text{time in seconds}}$$

- *Directly.* This method requires a long stretch of space, at least 150 m, without echoes e.g. the playing field. You will need to be very explicit with instructions, as you can't talk to pupils once they are that far away. You really have to use a starting pistol for this. Pupils need to time from seeing the smoke as the pistol is fired to hearing the bang. Only *you* can fire the gun, pupils should be involved in timing and measuring the distance. Speed is calculated as distance divided by time.

Interference

A signal generator, loudspeakers and microphone can be used to demonstrate interference. If you use two fairly large speakers facing each other, pupils will be able to hear the interference (loud and quiet regions) as they walk between the two speakers. If you pick up the result with a microphone and feed it to a CRO a qualitative investigation varying frequency or distance apart of speakers can be done. Bring out the parallels with interference between light waves. There is a fuller explanation for waves in general in Chapter 11.

Transverse waves

These can be introduced here or when work on ripple tanks is carried out (Chapter 11). If here, a good start is with a Mexican wave. Look for the direction of the wave and compare with the direction of each person (particle) involved. You can also do this with the Slinky, but this time move it from side to side. The pulse will be reflected as it is with longitudinal waves. Setting up a wave shows that particles individually move at right angles to the direction in which the wave travels. If you get the right frequency you can set up a standing wave with fixed stationary points (nodes) and maximum movement regions (antinodes).

Ultrasound

See either of the videos mentioned in Resources. Also you can refer to the motion sensor described in Chapter 5.

Assessing pupils' learning

- Give numerical practice in finding distances by sonar.
- If your school has a decibel meter use it in a group investigation into sound proofing. An alarm clock could be your source of sound and you

can brainstorm ideas for reducing the loudness. Research into noise pollution and ways of reducing it, write a leaflet of recommendations.
- Prepare a wall chart showing a use of ultrasound, try to ensure there are as many different ones as possible for an interesting display.

ENRICHMENT AND EXTENSION ACTIVITIES

Resonance is an interesting phenomenon with varied examples. This can be observed when vibrations are transmitted from one part of a system to another. A nice demonstration called Barton's Pendulums is pictured in Figure 10.1. The

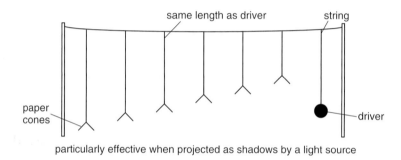

same length as driver   string

paper cones

driver

particularly effective when projected as shadows by a light source

Figure 10.1 *Barton's Pendulums.*

pendulum picking up the maximum energy from the driver is the one that has the same length. It has the same *natural* frequency as the driver i.e. frequency that it adopts when left to swing freely. The behaviour of the different length pendulums is also interesting. They are forced initially to gain energy and swing but since the frequency does not match their natural frequency they lose the energy quite quickly, passing it back to the driver. The demonstration is most effective if the pendulum bobs are white cones of paper and it is held in a darkened room with a bright light shining along it to throw its shadow on the wall.

Examples of resonance include:

1 The collapse of the Tacoma Narrows bridge whose concrete roadway began to vibrate with a particular wind speed until it fell into the river below.

2 The shattering of wine glasses by powerful sopranos. In this case the note sung would have to have the natural frequency of the wine glass. However, although this is believed to have been done, in practice singers would have a better chance if they used amplification.

3 The breaking of bridges when marched on by troops in regular step. If the frequency of the steps matches the natural frequency of a part of the bridge it can vibrate vigorously and break. For this reason troops break step when crossing.

RESOURCES

Video

- *Sound*, a Science in Action programme from the BBC.
- *Sound* a Channel 4 Schools Broadcasts.
- Scientific Eye *Hearing and Sound* and *Hearing the Sound* from Yorkshire Television.

Teacher resource

- Some interesting ways to model the ear are described in the notes in the March 1999 edition of School Science Review, Volume 80, Number 292.

chapter 11

Characteristics of waves

BACKGROUND

The basic principles underlying the way we interpret trains of waves and their behaviour at have been in place for more than three hundred years. They were put forward by a Dutch scientist, Christian Huygens. He regarded a wave front as a line joining points which were in 'phase' that is, were all at their crest or trough or any point provided it was identical on every wave. In this way the wave front on the coast would be parallel to the beach, ignoring currents, whereas the wave front caused by a point source, such as a pebble in a pool, would be a circle. A wave front arriving at a gap roughly the size of one wavelength is considered to effectively start out again with new waves.

The study of radiation and its value in our world of rapid communication is relevant to pupils. A ray is a way of looking at a train of waves. As waves, on the ocean say, travel forward their direction could be represented by a line with an arrow i.e. a ray. For radiation such as light, with a very small length of wave, we tend to draw the ray only. For longer waves, such as sound, we tend to draw the wave front.

The only wave which has received much attention in Key Stage 2 and 3 is the sound wave. The links between sound and all the other waves pupils will meet are:

1 The familiar shape (sine wave) they have seen on the CRO.

2 The properties they have in common (reflection, refraction, diffraction and interference).

Using ripples, which are transverse waves, these properties of waves can be clearly demonstrated.

There is, however an important distinction in the type of wave. Sound waves are called longitudinal not transverse. Longitudinal waves can not be transmitted through space since vibrations are passed by particles of a medium, but some transverse waves can.

The electromagnetic spectrum is an abstract concept which pupils struggle with even though the properties of the various members are well-known to them. Try to ensure that the different sections are seen as part of a complete spectrum extending in both directions from visible light, bringing out the similarities between them despite their varying wavelengths. It's also worth highlighting that not all radiation is bad radiation and, in fact, all our energy is received from the Sun in the form of radiation.

Common misunderstandings
- radiation is harmful
- water travels along with a wave, a misconception not helped by the fact that tides come in

KEY STAGE 4 CONCEPTS

Pupils should be taught:

- *That waves transfer energy without transferring matter.* Whether a wave is longitudinal or transverse, it transfers energy from source to receiver. The medium in which the wave travels moves, but does not have resultant movement in the direction of the wave. This deserves a mention each time a wave is encountered. You can show that energy is received whenever a wave arrives:

    Examples: *Kinetic energy at the far end of a Slinky.*
    Kinetic energy transferred to electrical energy at a microphone.
    A wave reaching your sand castle.
    Temperature rise where infra-red radiation hits the skin.

    Each part of the medium receives energy but it is passed on.
- *About longitudinal and transverse waves in ropes, springs and water.* See also Chapter 10. Ropes can be used to show transverse waves. They are not as versatile as the Slinky but are more effective as a satisfying crack demonstrates that energy has been transferred to the far end. A pulse (just one oscillation) is clearly seen. Pupils can stand at nodes and antinodes if you set up a standing wave. Then they can find the wavelength of the wave between two alternate nodes, not adjacent ones.

    Ripples in water are transverse which you can show in a large trough. However the best way to show other properties of ripples is a purpose built piece of equipment, the ripple tank. I would rather have one really good demonstration kit than a class set, firstly because they are tricky to set up, secondly they are difficult to interpret unless directed and thirdly have enormous potential for 'accidents'. Details for setting up are given in the Activities.

- *That waves can be reflected, refracted and diffracted.* The ripple tank can be used to bring out the equivalence of the angles of incidence and reflection, relate change in wavelength to change in speed in a different medium (it is in the shallow region that the waves travel more slowly), and show how the amount of diffraction is dependent on the width of the gap relative to the wavelength. It will also show the interference between two sets of waves that start initially from the same source.

- *The meaning of frequency, wavelength and amplitude of a wave.* See any text to see how wavelength and amplitude can be shown on a diagram. Ask students to imagine so many crests passing in a second – this is the frequency. Check that they can tell you what units are appropriate for each – wavelength is in metres, amplitude is in metres, frequency is in number/second or cycles/second and one cycle per second is given the name hertz (Hz).

- *The quantitative relationship between the speed, frequency and wavelength of a wave.* The speed of a wave is greater if there are more waves per second and if each wave is longer.

$$\text{speed} = \text{frequency} \times \text{wavelength}$$

$$\frac{m}{s} = \frac{number}{s} \times m$$

A quantity that is just a number has no units.

- *That longitudinal and transverse waves are transmitted through the Earth, producing wave records that provide evidence for the Earth's layered structure.* Waves which transfer energy during earthquakes are referred to as seismic. They are identified as three distinct types, primary (P), secondary (S) and long (L). The P waves are longitudinal and travel through the mantle and the core of the Earth, travelling faster where rocks are denser. The S waves are transverse and do not travel through the dense core of the Earth. The L waves are transverse and travel only in the surface causing more damage than P or S. By studying the time of travel of such waves, using a seismograph, estimates of densities of rocks in the Earth's structure can be made. It can also be used to help determine the liquid/solid ratio.

- *That the electromagnetic spectrum includes radio waves, microwaves, infra-red, visible light, ultraviolet waves, X-rays and gamma rays.* The part of the spectrum that we call a rainbow is a very small part of a much more extended range of frequencies. We can show that some rays exist beyond the red light that our eyes can detect. Since red light has the longest waves of the visible section, we must assume that the next along, or infra-red, have an even longer wavelength. We can also detect waves at the other end of the spectrum (ultra-violet) which have shorter waves than those at the visible blue section. And why should the spectrum stop then? Indeed it does not and, although we can not see any more, we can detect their effects and make use of them. Standard texts all carry diagrams of this spectrum. Although it is continuous, various sections carry names reflecting their discovery or history. Pupils need to know these names and should be able to compare and contrast them.

Points in common:

1 They are all electromagnetic radiation. This means their mode of travel is by vibration of electric and magnetic fields, no material is required to carry them (unlike sound waves).

2 They all travel at the same speed in a vacuum (space). The speed is the same as that given in Chapter 9 for light i.e. 300 000 000 m/s

3 They all show the characteristics of waves, i.e. reflection, refraction, diffraction and interference and the latter become increasingly difficult to detect as wavelengths become smaller.

4 Except for those with the longest wavelengths, they all react to darken photographic film.

Points in contrast:

1 They vary in wavelength from the longest (radio waves at several kilometres) to the shortest (gamma rays of the order of 1/100 000 000 000 m). The same range increases from low to high frequency since all travel at the same speed.

2 Since energy is transferred some can be considered dangerous, particularly those at high frequency.

3 They have different uses depending on the way in which they interact with matter and these can be dealt with as each type of wave is encountered.

- *Some uses and dangers of microwaves, infra-red and ultraviolet in domestic situations.*
- *Some uses of radio waves, microwaves and infra-red in communications.*
- *Some uses of X-rays and gamma-rays in medicine.* Uses and dangers are well documented but these points might help clarify some facts.

Microwaves in the domestic oven are produced so that their frequency is the same as the frequency of vibration of a water molecule. The action of resonance (Chapter 10) causes the molecules to vibrate rapidly, producing the same effect as heating the food i.e. raising its temperature. There are safety implications here and ovens are designed to be impossible to operate unless closed. It is quite possible to cause an explosion however so discuss the dangers of cooking anything sealed.

The **Doppler effect** can be detected in any wave motion but an application of which pupils may have heard is using microwaves in speed traps. The effect is as follows. A wave emitted by a stationary source is received at the same frequency by a stationary observer. However the movement of either source or observer changes the received frequency. This can be illustrated by the effect of a passing fast car. You hear the pitch of the sound drop as the vehicle passes. Figure 11.1 below explains the effect.

A device used as a speed trap emits microwaves at a certain frequency and receives an echo from an approaching car. The change in frequency is used to calculate the speed of the car.

Low power infra-red, as found in remote controls, presents no dangers but since any hot object emits infra-red radiation there are clearly dangers from irons, grills etc.

Ultra-violet radiation is that part of the Sun's spectrum which can cause cancer of the skin in spite of its giving what is normally referred to as a healthy tan.

Microwaves became important in the Second World War as radar, conveying information about nearby aircraft etc. It has now developed into a vital part of modern communications as it carries signals to and from satellites orbiting the Earth. At these wavelengths they do not appear to represent a danger.

As well as its use in remote control devices, infrared radiation is used to

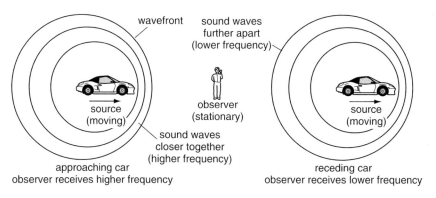

Figure 11.1

convey pictures when visible light is not available, for instance, heat-seeking cameras searching for survivors in collapsed buildings and tumour mapping.

X-rays are man-made in a vacuum tube and generally have only a small range of wavelengths. They are used diagnostically since they penetrate flesh more easily than bone. Gamma rays are emitted by some radioactive elements and can have a range of energies. They can be used diagnostically but are more commonly known for their use in the killing of cancerous cells.

KEY STAGE 4 ACTIVITIES

What do waves transfer?

If you begin with the Mexican wave or the water wave or a sound wave, in each case ask

- *What moves along?* A wave.
- *What actually moves?* People, water, molecules.
- *Which way do individual people or particles move?* Various descriptions but resulting in a vibration around an equilibrium point.
- *What actually arrives at the far end?* A message, energy to smash rocks, energy to vibrate an eardrum.

Ripples

A ripple tank should come with a box of gadgets and a class set of hand held stroboscope discs. It should be nice and clean to begin with. Set it up so that it is steady and reasonably level. Remember to put the plug in the base before you add any water. This should not be too deep, about 2 cm. The ripples are illuminated in different ways, either with a lamp shining from beneath on to the ceiling, or from on top onto the table. Either works well. You can make ripples by dipping in your finger. Pupils can see that they are circular and watch them reflecting from straight and curved barriers which you set up in the water.

To make a continuous train of ripples or wave a small motor is provided attached to a wooden rod. This arrangement is suspended on rubber bands so that the rod just touches the water. An offset device makes the motor, and therefore the rod, vibrate at a speed which is determined by the voltage across the motor. This creates a succession of waves of fixed frequency.

At this frequency the waves can be seen better if 'frozen' by viewing through a disc with slots in to give a stroboscopic effect. You hold the disc with its handle facing you, then spin it by placing your finger through from the *front*. Tell pupils to experiment with the speed of spinning until the ripples appear stationary.

Once you have established your plane waves show the characteristics of waves as follows:

- Reflection: place a plane barrier in the water at roughly 45° to the direction and observe them reflected through 90°. Try other angles.
- Refraction: place a piece of glass or Perspex flat down in the water so that the water above it is shallower than the water elsewhere. You will observe that the waves are noticeably shorter in the shallow water, i.e. the wave is travelling more slowly. Also they are turned through an angle as they travel over the plate – this will need careful inspection to convince pupils.
- Diffraction: allow the waves to pass through a gap about as wide as one wavelength. New waves spreading out from the gap will be seen, try the effect of varying the gap.
- Interference: use three barriers to make two gaps and note the fan pattern of maximum and minimum disturbance. Compare these with the patterns of interference identified with other types of wave.

Seismic waves

A model of a seismograph can be made from a flexible strip e.g. hack-saw blade with an inked paint brush attached. Set this up so that the brush touches a strip of paper stuck on a trolley. Nudge the blade as in an earthquake and give the trolley a gentle push to get a seismogram drawn out. If you want to produce a realistic graph you will need to control the paintbrush to give very gentle waves at first (P), then larger ones (S) and finally the damaging L waves.

The electromagnetic spectrum

Obtain a visible spectrum as explained in Chapter 9 and ask why the spectrum stops at the red and blue ends. Discussion should result in the idea that it is our inability to see other wavelengths/frequencies. It can be likened to our inability to hear audio frequencies that can be heard by bats. Direct evidence for the extension of the spectrum is perhaps not too convincing but should be mentioned. Beyond the red end the radiation can be collected by a photodiode sensitive to infra-red radiation. The signal can be detected as a small voltage when the diode is pointed past the end of the red section.

At the blue end, ultra-violet radiation will cause fluorescent materials to glow, either some bright white fabric washed in a strong solution of detergent with whiteners or some fluorescent paper. You can of course detect radio waves by tuning in a receiver, but from visible light to radio waves is too big a leap of the imagination for most of us to make. Various resources will help in the understanding of wave motion and they are listed overleaf.

ENRICHMENT AND EXTENSION ACTIVITIES

There is a good opportunity for some really interesting display material here with as many aspects of the spectrum as possible being depicted. Examples include: sources of radiation, methods of detection, uses, dangers, characteristics with a rainbow at the centre for the visible section.

For able pupils the study of the transmission of radio waves can be developed by looking at carrier waves and modulation. This is dealt with well, and at a suitable level, in *Physics Matters*.

Pupils can study oscillating bodies as the source of waves. Examples include pendulums and oscillating springs with a small mass attached. The topic could be extended with research into some simple timing devices.

Some students should consider the contrasting evidence for the nature of light. Diffraction and interference are the properties of waves, yet a phenomenon called the photoelectric effect points to its particle-like character. You can find more details below.

RESOURCES

Video
- *Electromagnetic spectrum* is on a Physics in Action tape available on loan from the Institute of Physics, Education Department, 76 Portland Place, London W1N 3DH.
- Posters, which would be useful for pupil research are available in the *At A Glance* series from The National Radiological Protection Board, Chilton, Didcot, Oxon, OX11 0RQ Tel 01235 831600.

Information technology

- A good resource for the whole range of waves is a CD-ROM called *Waves and Vibrations* from Anglia Multimedia. It is available on approval and comes with Activity Sheets. Contact them at PO Box 18 Benfleet, Essex SS7 1AZ, Tel 01268 755811 or see their Web Site at www.anglia.co.uk

Teacher's resources

- *Physics Matters* by Nick England published by Hodder and Stoughton.
- For a simple treatment of the wave particle duality see *GCSE Physics* by Tom Duncan published by John Murray.

4 The Earth and Beyond

chapter 12

The universe today

BACKGROUND

This is a topic where it is fine to let pupils get a bit carried away, even if some of the information they come up with is a bit more Star Wars than astronomy. The approach can be largely through research and project work. This should be as pupil-led as possible but should have a structure so that important concepts and demonstrations are included. Astronomy is a popular topic because it is intrinsically interesting and provides opportunity for imaginative work on space travel and aliens.

Some demonstrations using model solar and star systems will be necessary to introduce and clarify ideas. They will help pupils to appreciate the difficulties encountered by early astronomers in trying to predict the movements of heavenly bodies. Although it was known that some bodies moved *relatively* quickly, in general the stars seemed to be placed on a canopy which floated high above the Earth. The telescope was not invented until early in the seventeenth century. Galileo first used it to increase our knowledge of the planets and to distinguish them from stars. He observed the moons of Jupiter. It was the work of Newton which established the concept of a force called gravity which was keeping the planets in their orbits around the Sun and the moons in their orbits around 'their' planet.

In Key Stage 3 most youngsters will already be able to tell you that the Earth spins and that it also goes around the Sun, though it is likely that there will be confusion over how long each of these events takes and what its effects are on our lives. Any models you decide to use must come with the proviso that distances can not be represented as they are so vast. Practical work necessarily consists mainly of models and daytime exploration tools but there are good videos and CD-ROMS available which effectively backup mechanical models. A good set of slides of the planets is useful too.

You will often find pupils who are really keen and knowledgeable in this area, so spread their expertise in group work.

Common misunderstandings

- completely wrong ideas about scale, either of sizes or distances
- completely wrong ideas about eclipses
- seasons change because the Earth gets nearer to/further from the Sun
- confusion between stars and planets
- pupils get many wrong messages from film and TV images, such as space is full of stars and planets, particularly planets containing life
- there is no gravity on the Moon

KEY STAGE 3 CONCEPTS

Pupils should be taught:

- *That the apparent daily and annual movement of the Sun and other stars is caused by the movement of the Earth.* It is part of our every day language to use phrases such as 'the Sun sinks slowly in the West' and it is hard to eradicate the idea that it is the Sun that moves. Relative to the solar system, the Sun is fixed and the rotation of the Earth on its axis means that sometimes we face towards it (day) and sometimes away (night). Clearly this happens every 24 hours. At the same time the Earth goes around the Sun in one plane, a distance of 950 million km. This takes one year. If the Earth's axis of spin were at right angles to this plane all days would be the same in terms of hours of daylight and, therefore, seasons. However it is tilted at 23.5° so in our winter the Northern end of the axis is tilted away from the Sun, in our summer the Southern end is tilted away from the Sun.

S Warn pupils that studying the Sun cannot be done directly even through coloured or smoked glass without *danger to the eyes*. The only safe way is to throw an image of the Sun onto a screen.

Because the Earth's axis is at a constant angle to the plane of its orbit the same stars are always seen overhead in the Northern hemisphere. A camera on open shutter, left in a fixed place pointing at the sky for several hours at night, shows star streaks, which are good evidence for Earth's spin. The pattern of stars not directly overhead changes throughout the year as Earth moves round the Sun. Watch out for photographs in various magazines.

- *The relative positions of the Earth, Sun and planets in the solar system.* Any type of model solar system will give a picture and help to explain why it was so difficult for early astronomers with little technology to work out just what was going on. Each planet has its own time of orbit and so we see them at what seem to be random times, although of course careful study over a long period of time enables their course to be entirely predicted.

- *That gravitational forces determine the movements of planets around the Sun.* In Chapter 7 we looked at the pulling force called gravity. Newton proposed that this force acts between all objects which have any mass. It is dependent on the mass of the two objects and so is not apparent for items of small mass. For objects as big as stars, planets and moons, however, it provides a strong attraction, which keeps them moving in their orbits. Newton's First Law describes how, in the absence of unbalanced forces, moving objects continue to move with the same speed in a straight line. This is what planets and moons would do but for the pull of gravity accelerating them continuously towards the nearest massive body. Explain it in terms of a ball on a string being swung around your head. The tension in the string represents the pull of gravity and, if it should be suddenly cut off, the ball would fly off at a tangent to the circle not away from the centre.

  Gravity pulls the moons to the planets, the planets to the stars, the stars together in groups called galaxies and the whole Universe together. Nevertheless the energy that these larger bodies has means that they are still all moving apart at high speed but gravity may yet win and succeed in pulling them all back together again.

- *That the Sun and other stars are light sources and that the planets and other bodies are seen by reflected light.* It is important for pupils to understand that only stars are hot enough bodies to produce their own light. All other objects, planets, comets, asteroids, moons are only seen when reflecting a star's light. It may be necessary to emphasise that the Sun is a fairly average star as stars go, we only refer to it as the Sun because it is our 'local'. Studying the phases of the Moon helps to support the idea of reflected light as only the side illuminated by the Sun can be seen.

- *That artificial satellites can be used to observe the Earth and to explore the solar system.* Satellites are put into orbit for many reasons; communications, observation of varying kinds and the study of space away from the light pollution almost inescapable on Earth. Their orbiting path is maintained by gravity in the same way as the Moon's, in fact the Moon is described as a natural satellite. From an orbiting satellite, resolution almost down to the details of the human face is possible.

KEY STAGE 3 ACTIVITIES

The spinning Earth

It is a good idea to find out what pupils already know by having some true/false questions to try in groups. If they are a bit hazy about the formation of night and day then let each pupil pretend to be Earth. In a darkened room they spin slowly anticlockwise in the presence of a fixed light such as a projector.

- *When is it night, when is it day?* It is clear that sometimes the Earth faces the Sun, sometimes it faces away.
- *How does the Sun appear to move if you think of yourself as standing still?* The Sun moves across the sky in a horizontal line.
- *The Earth does not quite spin upright. Can you find out the real path of the Sun?* An arc going East to West.

The height of the arc is greater in summer than in winter and this is another consequence of the angle of the Earth's axis of spin. It's not easy to replicate this with a pupil so resort to a globe. The bigger it is the better (or even use a large beachball with a few countries drawn on). The Sun will need to shine so a bright bulb will be necessary. Show pupils where they are on the globe and tilt it at about 25° with the northern hemisphere pointing away from the Sun. As it spins show how the day is short and night long in the UK, i.e. it is winter. Now move around the Sun a quarter of an orbit at a time anticlockwise. If you keep the Earth's angle of tilt pointing to the same corner of the room you will pass through Spring, Summer and Autumn. Measuring the angle of inclination of the Sun at times during the day can be found using a simply made device and if dealt with graphically illustrates the point well. A clinometer is a useful instrument but it does need a very bright day to be effective. Some data you take in yourself from data tables probably available in your library may give more reliable graphs.

Sundials are good to make and can be as simple as a stick in the ground where you can mark the hours, ensuring that the stick is upright. Pupils may be ready for a more accurate device and can be given a design. (There is a template to be found in the Nuffield Science for Key Stage 3) The stick whose shadow provides the marker is called a gnomon and must be carefully placed for accuracy. Note that the hours are not evenly spaced – it's interesting to compare with an inexpensive garden ornament that often has evenly spaced markings, which would not work. They will find when they use their sundials that they must read the time appropriate to the latitude and also have to adjust for British Summertime.

Pictures of star groups are useful to show the patterns pupils are most likely to be able to see. They will know that various constellations are linked with star signs and will expect that their sign is visible in the sky during their birthday month. In fact 'their' constellation will be visible at the opposite end of the year, for instance Gemini (zodiac sign May to June) can be seen in the winter sky (November to December). Convert star maps by tracing from newspapers onto OHTs. Seeing how the night sky gradually changing with the months links with the movement of the Earth around the Sun. Add overlays with lines that show up the best known constellations. When you remove them it is easy to see why they can not readily be identified. Emphasise the pole star, which is always directly overhead.

Solar system

Assemble a set of spheres to show some idea of the scale of the sizes of the planets. The relative sizes are approximately as shown below.

Planet	Approx. diameter (km)	Scale diameter (mm)	Model
Mercury	5 000	5	Ball bearing, peppercorn
Venus	12 000	12	Small marble, bead
Mars	7 000	7	Ball bearing, dry pea
Jupiter	143 000	143	Child's football, round melon
Saturn	120 000	120	Grapefruit
Uranus	51 000	51	Apple, ball
Neptune	50 000	50	Apple, ball
Pluto	2 300	2.3	Mustard/cress seed

A good mnemonic for the nine planets is 'My Very Easy Method Just Speeds Up Naming Planets'.

A device that shows the relative movement of the planets, from a simple wind up model to a CD-ROM, is called an orrerry. See Resources for pupils who want to extend their interest in this topic.

Gravity

This should be treated very simply at Key Stage 3. Aim for pupils to be able to explain why people do not fall off the bottom of the Earth, in other words that the word down means towards the centre of the Earth. You can do this with Lego people and a globe. If the idea goes down well take some of your people to the Moon.

- *Why do they stay on the Moon?* Pull of gravity (which is less than on Earth).
- *Why are they pulled to the Moon and not the Earth?* They are nearer to the Moon
- *What keeps the Moon going around the Earth?* This will be very difficult to answer. Encourage answers that suggest the Earth is pulling on the Moon. Otherwise it would go travelling off at a tangent. Even more encouragement for ideas about this being the same force that keeps the people on the Earth and Moon.

Satellites

Ask pupils to think up reasons for launching satellites. They will have lots of ideas including geography, mapping, geology, communication, weather forecasting, spying and space observation and measurement. Give some idea of the scale of their distances with a crude model. If you represent the Earth by a globe, say 25 cm in diameter then the low-level satellites (weather, mapping etc.) would be roughly 1.7 cm above the surface of the Earth and the high-level ones (communication) at roughly 70 cm above. For comparison the Moon is at approximately 30 Earth diameters away so it should be placed at a distance of 7.5 m on this scale.

Satellites designed to study extra-terrestrial objects do not need to be in these orbits but are used simply to avoid the Earth's pollution, either chemical or radiation of various wavelengths. The data they receive is passed to computers on the ground to be translated into recognisable and useful images.

Assessing pupils' learning

- Explain, in writing, to your younger brother or sister why it gets dark at night.
- A postcard from a researched planet or moon – design it to look like a postcard and begin 'Dear Mum' or whoever is appropriate, describe scenery, food, leisure activities etc.
- Construct an alphabet/glossary of astronomical terms. Allot different letters to pupils and combine them for display.
- Describe life on Earth as seen by an alien who comes to call.

ENRICHMENT AND EXTENSION ACTIVITIES

A project on this topic gives plenty of scope and usually produces the type of pictorial outcomes that form excellent display material. Negotiate time in the resources base and IT rooms if appropriate. Organise a VIP (Year Head for instance) to present one or two awards. To give it structure, include at least half of the following.

- A helpful title.
- A stimulating visual aid, picture of astronaut, spaceship, etc.
- A list of aims, not just those from the science curriculum but IT skills, communication, presentation etc.
- Space for pupils' own ideas and organisation e.g. dates and deadlines.
- A broad and varied list of available resources, liaise with your Resources Head, IT specialist etc.
- A list of available items and demonstrations which can be used for the report back.
- Space for recording new vocabulary and its meaning – make it clear you are available to help here.
- Suggestions for ways through, perhaps a topic web arrived at by discussion in the group.
- A clear outline of the way in which you are expecting feedback – preferably by illustrated report to peers.
- Space to write a list of any resources used which you did not supply. This will enable you to check up doubtful points as well as adding to your list of resources for next time.

KEY STAGE 4 CONCEPTS

Pupils should be taught:

- *The relative positions of the Earth, Moon, Sun, planets and other bodies in the Universe.* The positions of the Earth, Moon, Sun and planets will be a matter of reviewing Key Stage 3 work. Introduce the following as well:

  1 Comets – these orbit the Sun but spend the major part of their time in outer space as their orbit is very elliptical.

  2 Asteroids – these are pieces of space debris made up of rock or ice. There is a large band of asteroids situated between the orbits of Mars and Jupiter.

  3 Meteors – these are broken pieces of asteroids which collide with planets or moons, either burning up in the atmosphere or colliding with the planet's surface, causing craters.

4 Moons – most of the planets have moons, Saturn having the most with 21.

5 Milky Way – the group of stars or galaxy to which our star belongs.

6 Galaxy – there are millions of galaxies making up the Universe, many are spiral with a bulge at the centre of the spiral arms and continuously rotate.

- *That gravitational forces determine the movements of planets, moons, comets and satellites.* The notion of centripetal force was raised in Chapter 7. It is the force required to keep an object moving in a circular orbit and is greater for:

1 More massive objects.

2 For objects with smaller radius of orbit.

3 Faster moving objects.

It is a force directed towards the centre of the orbit. You can help the idea along with examples of circular motion and the centripetal force.

    Example: *Wall of death – centripetal force of wall pushing to the centre.*

                    Circling skater on a rope – centripetal tension in the rope pulling towards the centre.

For planets around stars (and moons around planets) this force is provided by gravity. The orbits of planets are not strictly circular but slightly elliptical. This ellipse is even more highly pronounced for comets which come from the edges of or outside the solar system and swing around the Sun, speeding up as they get closer then slowing down as they move away again. Satellites that remain in a fixed position above the Earth are described as geostationary (see further details in Activities). Satellites passing over the poles are called polar.

Type of satellite	Low-level	High-level
Height km	850	36 000
Position	Orbit passes over the poles	Over the equator
Time for 1 orbit	100 minutes	24 hours
Function	Observation relaying pictures to ground covering whole Earth in 24 hours as it spins, pupils will be able to find this on the Internet	Observation from stationary position, each satellite viewing about one third of Earth's surface

A demonstration of the tidal effects of the Moon's gravity provides some evidence to support the theory of gravitational force.

KEY STAGE 4 ACTIVITIES

Phases of the Moon

Teach this in the context of understanding that only bodies as hot as stars can emit enough light to be seen in space, other bodies can only be seen when they reflect the light of stars. This is why planets, which must exist around other stars, can not be seen and even now are only just beginning to be detected. Our star, the Sun, can be represented by an OHP in a darkened room. Alternatively paint the Moon half yellow and dispense with the Sun. Make sure the yellow side always points the same way. Have the Earth, a sphere or a globe, about four times the diameter of the ball used for the Moon.

Begin with the full Moon, where the Moon is the opposite side of the Earth to the Sun. Keep the Moon at a good distance (in reality 30 Earth diameters away) so it is not in the Earth's shadow. Bring the Moon to the side of the Earth. The same part of the Moon will be illuminated but make sure pupils see that only one quarter of it reflects light to Earth. The crucial thing is that pupils are looking *as if they are on Earth* not as viewers at the side. Continue to show all the phases (Figure 12.1).

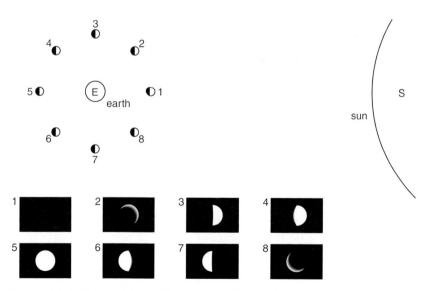

Figure 12.1 *Eight positions of the Moon are shown. The boxes show the view from the Earth.*

More able pupils will be able to appreciate that, because the Moon spins on its axis once every 28 days, we only ever see the same face. This aspect can not be brought out if you are using the painted ball method.

Planets

Key Stage 3 work on planets will need a review (see assignments).

Tables of data, available in astronomy books in your library, can be used in spreadsheets to find out relationships between variables such as density, mass, distance from the Sun, time of orbit, radius of planet etc.

Stars

Pupils can make a device called a planisphere which can give an instant picture of the stars that should be visible in the sky on any night in the year. A template should be provided (they are available from Hands-on Science Centres and Planetaria) together with a paper fastener as most pupils won't have one of these. The details of construction and how to use will be supplied with the template. You will need to run carefully through both because:

a) The construction involves fastening together two layers. The stars are drawn on a base layer. An offset cut-out section in clear film is pinned centrally over the base with a fastener so that it can be rotated revealing different sections of the sky.

b) In use the viewer must be standing facing in the correct direction, given on the template, and holding the planisphere overhead. The device must previously have been set to the right date and time. It sounds complicated but is quite accessible for most pupils with a little help.

Emphasise that the stars we see are at vastly different distances. Our nearest comprise our own galaxy, the group seen across the sky in very clear and dark conditions. This broad band is called the Milky Way. Good viewing is fairly infrequent due to pollution and bad weather but watch out for the weather forecast. The best viewing is in the winter, otherwise you have to wait until too late at night, so encourage pupils to wrap up well and stand outside to use the planisphere on a clear frosty evening. Other bright objects will also be seen, looking like stars. These will be planets and you can check what should be visible and in what part of the sky for the particular time of the year by looking in your newspaper.

Details of other space items, mentioned in the Concepts, can be reviewed by using the demonstrations described for Key Stage 3. To ensure that new words have been assimilated there is opportunity for team quizzes. Games such as

- name three properties of . . .
- design a space book for primary school children
- Pictionary with space objects
- a list of objects and characteristics to match up and stick in

Gravity

It is hard for pupils to understand why a force which acts in a straight line apparently produces circular motion. Try this argument.

- *What happens to a cannon ball if we fire it parallel to the Earth's surface?* It travels along, falling at the same time, until it hits the ground because the force is vertically downwards.
- *What happens if we fire it faster?* It goes further before hitting the ground.
- *What happens if, in our imagination, we fire even faster?* Eventually it will be so fast that it will leave the Earth altogether.
- *What if we choose the speed to be between one where it would fall back down and one where it would leave the Earth?* In this case we would go into orbit around the Earth (Figure 12.2), effectively falling continually. Achievement of the correct speed is essential to the successful launching of a satellite. Similarly the Moon is travelling at the right speed for its orbit. If it should be slowed down, by a collision with a large asteroid perhaps, then it would drop to a lower orbit around the Earth.
- *What force pulls objects so that they can not escape the Earth?* Gravity. But there is also an equal and opposite pull of the object on the Earth.
- *In which direction does it pull?* Allow 'down', then convert through questioning to 'towards the centre of the Earth'. A force that pulls towards the centre of a circle is called centripetal.
- Name some other objects orbiting because of gravity. Encourage as many answers as possible orbiting the Earth and the Sun.

Use a globe again to show how the gravitational pull of the Moon affects the tides. The oceans facing the Moon are pulled very slightly towards it. The bulge caused is a little delayed and follows the Moon as the Earth turns. On the side facing away from the Moon the water is furthest from the Moon so gets a little left behind hence a bulge on the opposite side causing two high tides per day

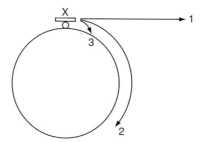

Figure 12.2 *Cannonball fired from X: 1. High speed – cannonball escapes. 2. Correct speed for orbit. 3. Low speed – cannonball falls to the ground.*

world wide. The Sun also has an effect, though it is smaller due to it being so far away. When Sun and Moon pull the same way the tide is extra high (spring tide) and when opposed the tide is low (neap tide).

Take the opportunity to discuss the Moon's gravity which is about one sixth of that on Earth due to its smaller mass. The implications of this are that, for the same energy, you can jump much higher and falling items will drop much more slowly. However, see Chapter 7 for the effects of air resistance.

Assessing pupils' learning

- Plot a Moon diary for a month showing a drawing of the phase whenever conditions allow. Complete a whole month but filling in the ones you could not see by studying the pattern carefully.
- Investigate the force needed to keep a body rotating in a circle. A suitable arrangement was suggested in Chapter 7. The radius of the rotating object can be varied and this can be related to the height of a satellite or moon. For safety this must be on a small scale, such as a bung on a string, and keeping to a limited number in a room.

ENRICHMENT AND EXTENSION ACTIVITIES

Take advantage of any astronomical events, such as an eclipse or comet, and keep up to date with which planets are visible at certain times of the year.

Visit a planetarium or Science Museum if possible.

Research historical ideas about gravity/weightlessness and report back via a flow chart or speech by a scientist/astronaut in the field (role play).

Emphasising the essential properties of astronomical systems can be explored by accounts of imaginative journeys to other solar systems where day/year lengths are different, gravity is greater/smaller, the number of moons is different etc. – there are lots of ideas to be gathered.

RESOURCES

Video

- *Understanding the Universe*, a series of videos from Yorkshire International Thompson Multimedia Ltd, contact them at 0161 627 4469 or through Nelson.
- They also have the Scientific Eye series for younger pupils including *The Solar System* and *Stars and Planets*.

Information technology
- Web sites such as NASA and the Royal Astronomical Society.
- A CD-ROM called *Interactive Space Encyclopaedia* from Andromeda Interactive Multimedia.

Teacher resources
- A good general resource is a book put together by the Education section of the Astronomical Association with the Association for Science Education simply called *Earth in Space*.

chapter 13

The evolution of the universe

BACKGROUND

This Chapter links with Chapter 12 and may overlap if treated through project work. It will not occupy much time in your work scheme but some pupils will want to discuss the theories and their implications. The facts are mind-boggling. There is a wealth of data available from CD-ROMS, books and the Internet. School libraries tend to like stocking up on such books as they look inviting and are much used, so you should find plenty of resources. If any pupils have problems reconciling theories with spiritual concepts they should know that these theories do not provide evidence for or against the existence of a God – no-one knows how, why or if any particular theory really happened.

The idea that a star has a life history was the product of the work of many astronomers over very many years, this is the nature of such long term processes. Yet startling events do occur to support the theory. One was the first observed supernova, an exploding star, in 1987. It happened in a neighbouring galaxy and produced more energy in a few seconds than the Sun will in its whole life time.

KEY STAGE 4 CONCEPTS

Pupils should be taught:

- *How stars evolve over a long time-scale.* Study of data from our own, and distant, galaxies has led scientists to believe that the Universe is an evolving system and that stars are 'born, grow up, grow old and die'. As with organisms the routes from birth to death can be various and the process is cyclical.

  The birth of a star is believed to take place in nebulae or clouds of gas. Hydrogen atoms are pulled together by the force of gravity in ever greater quantities. As collisions between them increase in number the energy rises. At a sufficiently high temperature (millions of degrees) fusion reactions take place between hydrogen ions, forming helium ions and releasing energy. The next stage depends on the quantity of matter which has collected.

Smaller stars, like the Sun, are described as yellow. After several thousand million years the hydrogen will all have fused and the star will expand to many times its volume, becoming a red giant. Eventually the gravitational pull on the particles causes the red giant to begin to shrink again until it reaches a dwarf state, a dead star with unbelievable density whose colour dies away as in a fire, white through red/brown to black.

Larger stars pass through a more dramatic life cycle as their cores build up to greater temperatures and pressures. This means that they explode over a comparatively short time and are called supernova. At these temperatures many nuclei fuse to form the heavier elements. The energy they have means that they fly apart as gas or dust which can then condense to form a new generation of stars and planets. The heavier elements on Earth, and in fact in your own body, have come from supernova explosions during the life of the Universe. On shrinking, the star may turn into a neutron star, another very dense body, which sends out strong pulses of electromagnetic waves. If the original star was even more massive it might shrink to a black hole. The gravitational pull of this is such that nothing can escape including light, hence it appears black. It can only be detected because of the complete absence of light within a certain halo, the limit of light's ability to escape.

- *About some ideas used to explain the evolution of the Universe into its present state.* The most commonly accepted current theory of the origin of the Universe is the Big Bang theory. In simple terms the Universe began as a point where all matter was concentrated and a huge explosion caused the matter to fly apart. Time and space began with this explosion. A recent estimate of the life of the current Universe is 12 billion years. The Sun was formed about 5 billion years ago. The initial bang caused unimaginable temperatures so particles created were very fast moving. As the cooling process began particles were drawn together by gravity ultimately forming planets and stars. Most stars grouped together in galaxies of around 10^{10} in number.

Evidence: *Distant galaxies are still moving apart from each other and us. This is revealed by measurements of 'red shift', a change in the wavelength of light reaching us from distant galaxies.*
Reverberations – detected in cosmic radiation.
Existence of heavy nuclei which require the temperatures of very hot stars for their formation and explosions to scatter them through the Universe.

The Steady State theory has little support now but was based on the idea that matter was continually being created and the Universe went on expanding indefinitely.

KEY STAGE 4 ACTIVITIES

Lives of the stars

The recommended CD-ROM is useful. Videos about astronomy usually illustrate the Big Bang well.

Prepare an illustrated sheet showing all the stages of star evolution and labels of information completely mixed up. Pupils cut and paste to make a poster that shows the stages in order. It will not be a straight chain of events since different stars evolve differently, mainly depending on their size.

History of the universe

Write the history of an atom of gold, which ends up in your earring. Discuss helpful ideas first. The atom could have been formed in a supernova and exploded far out into the space around. Once travelling it would continue until it experienced a force, perhaps a gravitational force pulling it towards a newly forming Sun. This stage could be repeated with the gold atom becoming a part of different stages of stars. Eventually it finds itself in the cooling gas cloud that becomes our solar system, and the part that becomes the Earth in particular. As the vapours cooled the gold atoms would form solids, probably buried in rock. Mining and forming the jewellery follows.

Assessing pupils' learning

- Choose a word from the list below and write a short report/Tomorrow's World type of presentation on it:

star, galaxy, pulsar, quasar, Hubble, black hole, fusion, dark matter

---

ENRICHMENT AND EXTENSION ACTIVITIES

Those who are confident in dealing with large numbers can find Hubble's Law and work out an estimate for the age of the Universe.

The red-shift effect mentioned above arises from an effect explained by Doppler and bearing his name, see Chapter 11. Some pupils can research this and its link with the position of spectral lines, which in turn provides us with information about the movement of distant galaxies.

RESOURCES

Video

- *Understanding the Universe*, a series of videos from Yorkshire International Thompson Multimedia Ltd, contact them at 0161 627 4469 or through Nelson.

Information technology

- CD-ROM – *Interactive Space Encyclopaedia* from Andromeda Interactive Multimedia.

Pupil activity

- To recommend – *The Time and Space of Uncle Albert* by Russell Stannard.

5 Energy Resources and Transfer

chapter 14

Energy transfer

BACKGROUND

Although energy is an everyday word it is not easy to give an explanation of what is meant by it. We know when we have got energy because we feel that we have the ability to get things done. In a similar way, things that have energy are able to get things done. When we do things, particularly those things we describe as energetic, we feel as if we are using up the energy. However we are not using up the energy, we are transferring it to something else. It is often transferred to air and our surroundings so it seems to have been destroyed.

A way of investigating energy transfer is to consider energy as taking different forms. In an activity a quantity of energy is transferred from one form to another. In this model, the energy forms described in the following table are included. In reality the distinction between them is artificial as you will sometimes find when you sort out energy transfers. They could be simplified as being of only two kinds, kinetic and potential, but this would demand great leaps for pupils.

Try to be consistent in the use of the word transfer. Always consider which type of energy has been put into an action and to which type it has been transferred during the action, the output. It is acceptable to use the terms useful energy transfer (heat energy from a grill) and wasted energy (heat energy and light leaving the grill and being dispersed in the atmosphere).

Work on energy resources can be done through project and research – it is

Type of energy	Further comment
Heat (thermal or internal) energy	Avoid 'heat' on its own except as a verb
Light	Distinguish visible light
Sound	A thunderclap
Electrical	Transferred by moving charge
Chemical	Transferred during chemical reactions e.g. combustion
Kinetic	Possessed by moving things
Potential (gravitational)	Energy of things moved upwards against gravity
Potential (Strain)	Energy of things under stress e.g. stretched rubber band
Nuclear	Transferred during nuclear reactions

important to discuss with pupils the limited supplies of fossil fuels and to try to raise their awareness of the crisis that will become a major cause for concern in their lifetime. Try to encourage thinking about how the crisis could be alleviated through:

- discovering new sources of fossil fuels
- researching new and more efficient ways of using renewable resources
- saving the resources we have

The remainder of the topic offers plenty of opportunity for practical work. Finding out about heat energy transfer enables pupils to apply their knowledge to preventing heat energy loss in various situations, from keeping soup warm to survival in extreme temperatures.

Common misunderstandings
- confusion between heat and temperature
- cold is a substance, rather like cold air ('shut the door and keep the cold out')
- heat rises
- when material expands, its molecules expand
- renewable means you can use something again

KEY STAGE 3 CONCEPTS

Pupils should be taught:

- *That there are a variety of energy resources, including oil, gas, coal, biomass, food, wind, waves and batteries, and that some of the Earth's energy resources are renewable and some are not. See the following table.*

Name	Method of use	Advantage	Disadvantage	Renewable
Oil	Burning in combustion engines	Can be extracted relatively easily	Pollution when burnt	No
Gas	Burning in power stations and for domestic use	Can be extracted relatively easily	Pollution when burnt	No
Coal	Burning	Can be extracted relatively easily	Pollution when burnt	No
Biomass	Burning organic waste, e.g.straw, dung	Uses waste product	Some pollution but not sulphur or lead containing	Yes
Food	Respiration	Takes place at body temperature	Not transferable from producer	Yes
Wind	Windmills	Free when available	Inconsistent, lots of windmills needed	Yes
Waves	Bobbing ducks or some type of piston	Free when available	Long lengths of coastline required	Yes
Batteries	Use where easy transport is needed e.g. camera, car	Convenient, easy to transport, may be re-chargeable	Impractical for supplying much power, costly	Yes
Hydroelectric	Waterwheel turns turbines	Free and constant	Only possible in sites where head of water sufficient	Yes
Nuclear	Nuclear reactions supply energy to produce steam	Clean, efficient	Expensive and risky after-care of fuel and plant	Not strictly, but in practice yes
Solar	As heat energy or conversion to electrical by cell	Efficient	Large areas of solar cells (panels) necessary so costly	Yes

- *That the Sun is the ultimate source of most of the Earth's energy resources.* Any one of the energy resources above can be shown to have derived its energy initially from the Sun, except nuclear energy. Radioactive materials became part of the Earth when the whole solar system was being formed so can not be said to have come from the Sun.

   Example: Oil is made up of minute sea animals, they derived their energy from the plant materials they ingested, these achieved energy through photosynthesis which effectively stores energy received from the Sun.

- *That electricity is generated using a variety of energy resources.* The principle of the dynamo or generator is dealt with in Chapter 4. A coil of wire must be rotated in a magnetic field in order to generate a potential difference between the two ends of the wire. The rotation can be achieved by applying a force from steam or a gas under pressure. The heat energy to make the steam, or pressurise the gas, comes from the energy source. The choice of source will be based on many factors, those in the list above and others based on location, local materials, pollution etc. The efficiency of power stations burning fossil fuels is greater than that of individual petrol engines so they are a better use of resources.

- *The distinction between temperature and the total energy contained in a body.* Temperature is a scale which tells us how hot something is. It is comparative i.e. a point of reference is chosen and other temperatures compared with it. On the Celsius/centigrade scale two points are fixed, 0°C and 100°C (freezing and boiling points of water) and the gap between them divided into 100 gaps called degrees. On that scale human body temperature is at 37.5°C, an object or environment which *feels* hot is close to or above this temperature.

   A body is hot because its particles are moving: the faster they move the higher its temperature. Some particles require more energy to raise their temperature by, say, 1°C than others so two objects of the same mass at the same temperature do not necessarily have the same amount of energy. We say their internal energy is different. The unit for energy is the joule (J). It is quite a small unit, lifting an apple by 1 m uses around 1 J of energy. Boiling a cup of water requires about 70 000 J. Compare these two examples.

   Example: *1 kg of copper needs 380 J of energy to raise its temperature by 1°C.*
   Its internal energy increases by 380 J.
   We feel it warm up or cool down quickly.

   Example: *1 kg of water needs 4200 J of energy to achieve the same temperature rise.*
   Its internal energy increases by 4200 J. If the energy is supplied by the same source as in the above example, the temperature rise will take longer to achieve.
   This also means that it has a lot of internal energy to release when

it is cooling so hot baths stay hot for an appreciable time. Conversely cold water will take a lot of your body heat if you fall in it.

- *That energy can be transferred and stored, that energy is conserved and that although energy is always conserved, it may be dissipated, reducing its availability as a resource.* Amounts of energy can not be measured directly. We can only measure what the energy achieves in terms of raised temperature, light emitted, load lifted etc. This means we are measuring only energy transferred in the activity, not the total energy stored in the body. With some energy transfers we detect energy lost to the surroundings, usually in the form of heat energy. Energy lost in this way can not be collected in again for use, it is described as dispersed. We must avoid saying that the energy has been used up. All the energy is transferred somewhere, the total energy of the system and its surroundings remains constant or is conserved.

You might feel that there is no energy crisis if total energy remains constant. However dispersed energy can not be recaptured so, although it is there, it is not available to be used.

KEY STAGE 3 ACTIVITIES

Earth's resources

This section is best covered as research/project work. Suggestions are set out in Chapter 12 for a structure to support pupils in consulting available resources and preparing reports for feedback to the group. These reports might, if time permits, be made by account, posters, models etc. and should include reference to the availability of the resource, its effect on the environment and whether or not it is renewable.

Energy from the Sun

Each group or pair can produce a cartoon-style flow chart to show how the energy of a source was derived ultimately from the Sun.

Generating electricity

A device that turns the coil in a magnetic field in order to produce a potential difference, is called a dynamo. On a small scale, you could turn the coil by hand (or get your hamster to do it!) and you can demonstrate this using a mounted dynamo. On a larger scale, as in a turbine, the coil can be turned by pressure. A cardboard replica set of vanes can turn your handle into a model, though non-functioning, turbine. Pupils can use their imagination to visualise the force of steam or gas turning the blades. The heat can be supplied by any of the fuels. Some pupils object that steam is not strong enough to do this but remind them of the power of the steam engine. The turbine can be turned directly by wind, waves, waterwheel etc. In either case the dynamo will show that the potential difference created depends on the speed of turning.

The solar panel operates in a completely different way. You may not have available a complete panel but can show the effect if your school has a small-scale equivalent attached to a tiny motor. In sunlight or bright lamplight the motor will spin, this will be more effective if you attach a colourful but lightweight propeller to the axle of the motor.

Temperature and heat energy

Hand out some long strips of paper and ask pupils to draw a giant thermometer. They pin it on the wall and then mark on it as many temperatures as they can, in order but probably not to scale. They might achieve a range from −180°C (liquid oxygen) to a million degrees in a star (the surface of the Sun is at several thousand degrees). Make a few resources available to help fill in at least ten fixed points between the top and the bottom, such as body temperature, freezing and boiling points of some substances, maximum in a desert, minimum at the South Pole etc.

There should not be too much emphasis on numerical work at Key Stage 3 but pupils should have an opportunity of supplying the same quantity of energy to two different materials and finding out for themselves that the temperatures do not rise together. This is most conveniently done with electrical immersion heaters since it is easy to use the same current for the same length of time with good accuracy. Warning: the heater may burn out if it is switched on without being placed in water or in contact with the aluminium block.

The items being heated should have the same mass and be under the same conditions as far as possible. Many school laboratories have purpose built aluminium blocks with a mass of 1 kg and slots for inserting the heater and a thermometer. These are commonly filled with oil to ensure good thermal contact so warn pupils not to tip them up. The rise in temperature of the block in a set time can be compared with the same arrangement and time for one litre of water. It is not really necessary to know how much energy has been supplied but it can be measured if an instrument called a joulemeter is available. It reads by flashing for every 100 J supplied.

Transfer and conservation of energy

Activity	Energy input	Energy output
Releasing clockwork toy	Potential energy of spring	Kinetic energy of toy
Pendulum knocking over wooden brick wall	Gravitational potential energy of ball	Kinetic energy of bricks
Brandy on Xmas pudding i.e. alcohol on Rocksil	Chemical energy	Heat and light energy
Battery operated bell	Electrical	Sound
Effect of light on silver bromide	Light energy	Chemical energy
Thermocouple and galvanometer	Heat energy	Electrical energy
Electrolysis of copper sulphate solution	Electrical energy	Chemical energy
Maracas or seed shaker	Kinetic energy	Sound energy
Toy car propelled up ramp	Kinetic energy	Gravitational potential energy

S These experiments work well as a circus provided that you keep an eye on the galvanometer. The alcohol burning must be demonstrated with these precautions emphasised.

- use safety goggles, staff and pupils
- remove the bottle of alcohol to a separate bench
- warm the alcohol in a water bath
- light with a long spill

In some of these you may find a string of energy transfers necessary but as far as possible just identify the input and the output, ignoring intermediate stages. You may want to add a fourth column indicating any energy loss pupils can suggest.

It is not a simple matter to demonstrate the Conservation of Energy under *any* circumstances so it is best taken as understood and referred to whenever energy transfers arise. A useful way of thinking about where the energy goes is in the use of Sankey diagrams. Figure 14.1 shows what they depict but note that the width represents a quantity of energy so the *total* width should be the same before and after.

Various groups concerned with energy saving will supply educational materials, see Resources.

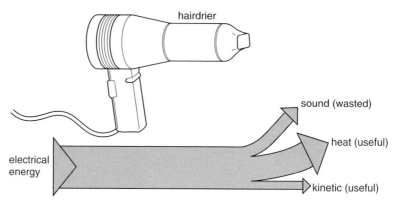

Figure 14.1

Assessing pupils' learning

- Put pupils in charge of generating electricity in a remote place; mountain, desert island, crisis zone etc. Ask for full details of how this is to be done. In reporting back the class can ask questions about pollution, reliability, etc.
- Practice with Sankey diagrams, e.g. car, firework, light bulb.
- Find out the energy supplied by different foods (written on the packet in kJ) and compare it with energy needed for a healthy life style (consult a biology book). Write a menu for an office worker or a weightlifter, after collecting data by questionnaire.
- Look at some appliances at home and decide where energy is wasted and where it goes.

ENRICHMENT AND EXTENSION ACTIVITIES

Debate issues involving the use and distribution of energy. They can become heated as environmental issues are raised. Try 'cars should be banned in towns' or 'by switching off unused lights we can help protect wildlife'. At Key Stage 3 pupils will need guidance, perhaps some hints cards, and debates should be kept short so that perhaps three different groups can debate in a lesson. It is tempting to provide your own solutions but resist it.

Ideas of particulate theory can be raised with pupils ready to consider these more abstract ideas. Some of the suggestions from the Key Stage 4 Activities can be used, leaving more time at the later stage for quantitative work.

KEY STAGE 4 CONCEPTS

Pupils should be taught:

- *That differences in temperature can lead to a transfer of energy and how energy is transferred by the movement of particles in conduction, convection and evaporation.* There are a number of experiments that nicely illustrate these methods of heat energy transfer. Their explanation should be in terms of the particulate theory of matter. Evidence for the theory, which is accessible to pupils, is provided by such phenomena as Brownian motion and diffusion (see Activities). Discussion and investigation result in pictures for solids, liquids and gases as found in standard texts. Use of the word particles at this stage helps to avoid any confusion about whether we are talking about molecules, atoms or ions.

  Conduction of heat energy is only really significant in metals. Describe the particles as being held together as if they were joined by tiny springs. When some, perhaps at the end of a rod, are heated they vibrate more vigorously and pull their nearest neighbours with them. These in turn make *their* neighbours vibrate and so the heat energy is passed along. The analogy is sometimes used that a parcel of energy is passed along from one person to the next but bear in mind that the part of the metal that the heat has passed along remains hot, i.e. it does not pass on all the energy as in a wave.

  Transference of heat energy by convection is significant in liquids and gases because the material needs to be able to flow. In this case heating causes the more freely moving particles to move apart slightly, meaning the region has slightly lower density. This region rises whereas the colder, denser material sinks to take its place. In this way circulating currents are set up. Expanding the analogy for conduction, it is as if the person who gets the energy moves forward with it pushing his/her neighbour along. This has a knock on effect until the person at the far end has no room so moves round to the front. If this continues, eventually the energy parcel will arrive at the far end still carried by the original holder.

  Examples: Heating element is placed at the bottom of a kettle or tank, the circulating currents ensure that all the water becomes hot.
  Thermals – rising columns of warm air on which birds and gliders gain lift.
  So-called radiators disperse a lot of their heat energy by convection, the currents rise to warm the top of the room but pull in cool air along the floor.

  The similarity between the words, conduction and convection, means pupils mix them up. I don't know of a very effective way to help here, but perhaps

171

associating ductile with metals and convector heaters with currents might help.

The process of evaporation means that the particles of a liquid which have the most energy, are the most likely to leave the surface of the liquid since they have sufficient energy to overcome the force of attraction of the remainder. As a whole then, the average energy of those left behind will be less than those that have left hence heat energy has been transferred. Evaporation is speeded up in these circumstances,

- warmer liquid, more particles have the energy to escape
- dry air, fewer particles of moisture in the air mean fewer particles returned to the liquid
- moving air means the fast particles get swept away making space for more

You could use these variables for an investigation into evaporation using different temperature liquids and hairdryers for moving air particles. The humidity is more difficult to control, the best way would be to put the samples in different places with a hygrometer that is read regularly.

- *How energy is transferred by radiation.* All waves transfer energy as discussed in Chapter 11. The higher the frequency of the wave, the more energy is transferred. The characteristics of electromagnetic waves mean that energy can be transferred across space, unlike conduction and convection. With reference to heat energy in particular, all hot bodies give off radiation, cool bodies receive it. The hotter a body the more high frequency radiation is emitted, hence the colour change from red to white as a piece of metal gets hotter. Going back to the analogy used above, the parcel is thrown from the first person to the receiver without touching anything en route.

  Different surfaces absorb and emit radiation at different rates, matt black surfaces being the most effective in both cases. This is sometimes thought to be confusing but just consider which is hotter, the emitter or the surroundings. If the emitter then a black surface will cool more quickly, if the surroundings then the same surface will warm more quickly. A shiny white surface does the reverse.

  Examples: *White houses in hot climates remain cooler (poorest absorber).*
  Shiny metal teapots keep tea hot for longer (poorest radiator).
  Solar panels are matt black (best absorber).

- *That insulation can reduce transfer of energy from hotter to colder objects, and how insulation is used in domestic contexts.* Pupils can investigate the effectiveness of insulation for themselves and can achieve reliable results particularly if datalogging techniques are used. The point that they may not appreciate is that air generally plays an important role in insulating materials. This is because air is a very poor conductor and, if it remains still, convection cannot take place. Movement of air can be reduced by confining the space such as in double-glazing, bubblewrap and expanded foam.

Example: *Ways of keeping the body warm for elderly people, skiers, etc.*
A cross-section of the material used for Arctic-style jackets shows the trapped air in insulating materials, breathable and permeable layers.

Example: *Ways of reducing heat loss from homes.*
These include loft insulation, floor coverings, cavity wall insulation, double-glazing, draught proofing, curtains.

KEY STAGE 4 ACTIVITIES

Particulate theory

The particulate theory needs reinforcing on any appropriate occasion. We have met it in sound, in properties of materials and now in the mechanism for heat energy transfer. It may be so familiar an idea that pupils are surprised to be offered evidence but, after discussion of dissolving and diffusion, they will like to see movement of particles.

First raise ideas of particles in solids, liquids and gases. Warm a lump of butter or lard over a water bath.

- *What do you see?* Melting.
- *Suppose it is made of particles, how would you describe what is happening?*
  Particles no longer strongly pulled together, fall under gravity
- *Imagine that we could heat a metal, where the particles are strongly held, to melting point.* The same sort of thing would happen.

Then dissolve a large crystal e.g. copper sulphate, in a beaker of warm water standing on an OHP.

- Imagine you have the ability to see tiny details like a very powerful microscope, what would you see? Clumps of particles falling off the crystal and breaking into still smaller pieces until they are so small that they are just seen as colour (in the case of salt they just disappear).

Light a scented candle and ask pupils to raise their hands when they can smell it.

- How is the scent reaching you? Particles are given energy to leave the candle and they mix with the particles of air until there are some in every part of the room (diffusion).

Tell pupils that they can now see the effect of air particles knocking some bigger particles around (Brownian motion). These are particles of smoke and they can just be seen with a good school microscope. You cannot really set this up in

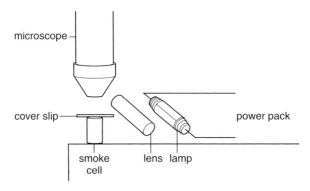

microscope

cover slip

power pack

smoke
cell

lens lamp

Figure 14.2

advance as the smoke can escape quite quickly. So take it slowly and explain what you are doing. Give pupils a diagram to complete and name while you are making sure it is working. The arrangement is shown in Figure 14.2.

The smoke is placed in a cell and illuminated by a bright light, usually part of the purchased equipment. The smoke is observed through the microscope and its tiny particles are seen to be in continuous random movement. The following will ensure success:

1 Switch on the bright light.

2 Focus approximately by finding the cover slip on the top of the cell and then lowering the lens a little.

3 Ensure that the cover slip is a good fit on the cell so that smoke doesn't escape too quickly. A little Vaseline rubbed carefully on the top of the cell might help.

4 Light a waxy straw and tip it flame-end away from the cell so that smoke pours in.

5 Quickly replace the cover slip.

6 Look through the eyepiece for tiny 'stars', which are continually jiggling about.

7 If you can not see it adjust the focus, make sure pupils know what they are expecting to see.

8 Don't despair if it proves elusive. Have something ready for pupils to get on with and try another set of equipment. It is well worth the risk to hear pupils' excitement when they see the movement. Ensure that it's understood that they are not seeing individual molecules but the comparatively large particles of smoke being pushed around by air molecules. I think of it as a football game where the ball can be seen but the players are invisible – the movement would appear random, at least to me.

Conduction

- *Which feels more cold, a metal bar or a wooden one?* Have both available for pupils check.
- *Why is that?* The heat energy travels easily through the metal away from the hand. A good confirming demonstration makes use of a rod, half metal (brass) and half wood, joined in the middle. Wrap paper, about 10 cm wide, tightly around the join and seal. You can then heat with a Bunsen burner at the join. The result is most effective if you keep turning the rod so that the paper chars without catching alight. You will get a clean line at the join showing black on the wood side and just a slight mellowing on the metal side. Clearly the metal is a much better conductor of heat.
- *Are all metals equally good?* This may be a class experiment or you may prefer a demonstration, as it can be messy. You need some equal sized rods of different materials, you will probably have brass, steel, aluminium and copper. Wearing goggles pupils should be given a little melted wax and 4 drawing pins so that they can attach a pin to each. Turning the rod so that the pin is on top stops it falling off while the wax is still soft. Once set, the rods can be placed on a tripod in a fan shape so that one end of each rests over a Bunsen. A metal tray must be used under the pins to catch wax. Pupils can time how long it takes for the heat energy to reach each pin, from lighting the flame to hearing the pin land in the tray. The results may not match the conductivities given in data tables due to uneven heating and the conduction of the tripod but it shows that metals vary quite a lot and reliably gives copper as the best.

A less messy alternative is listed in the resources together with suggested videos.

Convection

Two straightforward demonstrations are illustrated in Figure 14.3. Diagram A shows convection in water. To be effective you should use a large beaker to minimise the interference of hot tripod. Don't use a gauze mat and wait until the water in your beaker is nice and steady. The crystals of potassium VII manganate need to be as large as you can get, pick out a few with tweezers and drop them into the bottom of one side of the beaker down a straw. This must be removed very carefully without disturbing the water unduly. Begin to heat just under the crystals with a very small flame and you will see purple currents rising, moving across the top of the water and falling to make a complete cycle before it

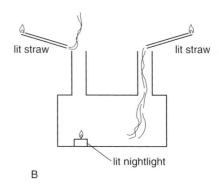

lit straw   lit straw

lit nightlight

A   B

Figure 14.3

all gets mixed up. A white background will help to make this visible from a distance.

This experiment can be repeated using sensors with datalogging equipment, without potassium(VII) manganate. Have one sensor situated low down, but not touching the beaker, and one in the top layer. If the logger has been set up for real time pupils can observe the two temperatures changing as the currents circulate.

Diagram B shows convection in air. Treat the glass chimneys with care as they are very fragile. Get everything else ready, including lighting the nightlight before putting the chimneys in place. The draught screen in front is necessary. Light a waxy straw and hold it flame upwards so that smoke pours down. Hold it first over the chimney above the candle, the smoke can be seen to be rising steadily upwards. Then move across to the second chimney. Immediately smoke will be sucked downwards. You can move from one to the other until your smoke runs out and it will be equally effective. Pupils will like to know that this technique was used to create currents of fresh air in mines by lighting fires under one of two shafts.

Evaporation

As this concept is a familiar one this could provide a suitable topic for a Sc. 1 investigation. Pupils can decide on their method of working, apparatus and controlling of variables.

Radiation

It is possible to feel the different rates of absorption of heat energy by attaching a piece of black polythene to the back of one hand and a similar piece of foil to the back of the other and exposing them equally to a heat source.

Small radiant heaters can be used but pupils must be warned not to attempt to touch or poke the element. A bright light bulb will do but must be mounted in an appropriate base as mains voltage will be used. Pupils should agree that the patch on the black side is warmer due to better absorption.

A piece of equipment known as a Leslie cube convincingly demonstrates which surface is the better radiator. The cube has four different sides, usually a matt black, a shiny black, a white and a shiny metal. The cube is placed on a heat proof mat and filled with very hot water. The palms of the hands placed at 1–2 cm away from the faces quickly reveal that the matt, black surface is emitting the most heat energy.

A device called a Crooke's radiometer is still to be found in some schools. It is like a miniature weather vane in a glass bulb. One side of each sail is painted black, the other shiny. In the presence of a source of heat the vanes rotate quickly in such a direction that shows more energy is being collected on the black side.

Insulation in use

Insulation can be investigated using containers of hot water, each with a thermometer, surrounded by different insulators. Pupils can decide how much water, and what other variables to control. It can be rather tedious as cooling is slow, but the data can be collected remotely using a datalogging package such as Insight or SoftLab. This will provide sensors to collect temperatures and a program to store them and plot them on a graph. You will be able to reset the axes to give expanded results, useful in this particular experiment. If different groups tackle different materials it will be possible to share data to gain a good range. The consensus should be that the most successful insulators are those containing air pockets or gaps.

Pupils can research the insulators used in the home. Its is also important to bring in costs of such measures compared with savings, not forgetting other benefits (such as sound insulation) and disadvantages (such as stale air and build up of bacteria).

Assessing pupils' learning

- Make leaflets for novice mountaineers or the elderly, making recommendations for keeping warm. While reference to all methods of transferring heat energy should be included, so should other survival tactics such as hot food, keeping dry, being active etc.
- Design a snakes and ladders game on energy conservation. For instance this could be based on a school where snakes show bad examples, like opening windows when the heating is on, and ladders show good ones, like switching off the lights when you leave a room.

• Show and provide a drawing of a vacuum flask. Ask how ways of transferring heat, conduction, convection, radiation and evaporation, are reduced as much as possible. How can it keep things hot *or* cold?

ENRICHMENT AND EXTENSION ACTIVITIES

Hand out large sketches of a house and set a challenge to attach pieces of fabric, polythene, foam, etc. to illustrate a well insulated house. Prizes for science involved and for presentation.

Ask pupils to explain the effect in Figure 14.4.

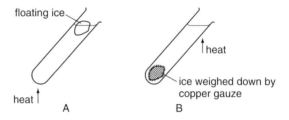

floating ice

heat

A

heat

ice weighed down by copper gauze

B

Figure 14.4

In Part A the ice melts quite quickly, in Part B the ice does not melt even though, if you are careful, you can boil the water at the top. The reason is that water is a poor conductor but in A, convection is taking place.

Datalogging programs can be used to find rates of cooling and compare them with the temperature of the cooling objects above room temperature. See Newton's Law of Cooling

Able pupils should use the equation for specific heat capacity.

> **heat energy transferred = mass × specific heating capacity × temperature rise**

Please note that heat energy transferred is equivalent to the change in internal energy. If energy is in joules, mass in kg and temperature rise in °C then you can show that the units of specific heating capacity are J/kg/°C. Simple examples are of the type described in the Concepts. More complex problems transfer heat energy from one substance, say hot metal, to one or more other substances, perhaps water in a specified container.

RESOURCES

Video
- A cartoon on the general topic of energy conservation called *Waste Watchers* can be hired from Leeds Animation Workshop, 45 Bayswater Row, Leeds LS8 5LF Tel 0113 248 4997.
- *Energy* from Channel 4 Schools, PO Box 100, Warwick, CV34 6TZ Tel 01926 436446.
- *Energy* is one of a group of science videos from Yorkshire International Thompson Multimedia at 0161 627 4469 or contact Nelsons.
- The same company has the Scientific Eye series for ages 11–14 with the same title.
- *Energy and Energy Transfer* from Granada Television.

Information technology
- CD-ROM *Understanding Energy* from Anglia Multimedia for KS3. Tel/fax 01268 755811 or contact SCA, PO Box 18, Benfleet, Essex SS7 1AZ.

Teacher resources
- Friends of the Earth will send a catalogue of their relevant educational materials if you contact them at FoE, 26–28 Underwood Street, London N1 7JQ.

Practical
- As an alternative to the conduction experiment described, thermochromic bars are a lot less messy and will last a long time. (Nicholl, Block 1 Nortonthorpe Mills, Scissett, Huddersfield HD8 9LA Tel 01484 860006).
- Datalogging packages include Softlab and Insight 2, a recent update.

chapter 15

Work and power

BACKGROUND

This Chapter is directed only at Key Stage 4 and is mainly concerned with quantitative work. Formulae are given for calculating transferred energy, its rate of transfer and the efficiency with which it can be transferred in various systems. Above all, it contains words that are defined carefully and must lose their everyday connotations in this context. There is practical work to be covered because the data should be collected from pupils' own investigations as far as possible. These are generally enjoyed as pupils like to measure their own power. Estimating their own efficiency is much more difficult so simpler machines than the human body are used for these measurements. Considerable work on energy resources will have been done at Key Stage 3 so a review of this will make a suitable starter for this section.

Common misunderstandings
- efficiency means how smoothly something works – while not entirely wrong this is not enough to define efficiency
- work – something only humans (and some animals under human direction) do, the more they do the more tired they become
- energy gets used up – hence the energy crisis
- confusion between global warming, greenhouse effect, holes in the ozone layer, acid rain and their causes and effects
- confusion of fuel and energy

Pupils should be taught:

- *The meaning of energy efficiency and the need for economical use of energy resources.* Efficiency is the proportion of useful energy transferred relative to the energy input. Useful energy here means achieving the desired result as opposed to dispersing it (wasting). It is normally expressed as a percentage:

$$\text{efficiency} = \frac{\text{useful energy output}}{\text{energy input}} \times 100$$

Example: A light bulb uses 100 J/s but only 40 J/s are converted into visible light energy.

$$\text{The efficiency of the bulb} = \frac{40}{100} \times 100$$
$$= 40\%$$

Note that there are no units for efficiency and it can't have a value of more than 100. For ways of calculating energy input and energy output see the next section.

- *The quantitative relationship between force and work.* Work is done whenever a force moves. Since work has been done *on* something, *its* internal energy has increased by that amount of work. Similarly the worker will have lost energy equivalent to the work it did. *Work done is equivalent to energy transferred.*

The bigger the force and the further it moves the greater the work done.

work done or energy transferred = force × distance moved

$$J = N \times m$$

If energy is lost to the surroundings during an energy transfer, the energy transferred from the worker (energy input to activity) will be more than the energy transferred to the object (useful energy output).

Example: Jill uses a force of 50 N on a car jack to raise a car weighing 1500 N. If she has to move the handle of the jack a total distance of 5 m to raise the car by 0.1 m what is the efficiency of the jack?

$$\text{energy input} = 50 \times 5 = 250 \text{ J}$$
$$\text{energy output} = 1500 \times 0.1 = 150 \text{ J}$$
$$\text{efficiency of jack} = \frac{150 \times 100}{250}$$
$$= 60\%$$

To calculate power in terms of the rate of working or of transferring energy. Power is the rate at which the work is being done or the energy being transferred.

$$\text{power} = \frac{\text{energy transferred}}{\text{time taken}}$$

$$\text{watts} = \text{joules/seconds}$$

Example: Terry lifts 200 cans, weighing 5 N each on to a shelf 1.5 m high, in 5 minutes. Calculate the worker's power.

$$\text{Energy transferred} = 5 \times 1.5 \times 200 \text{ J}$$
$$= 1500 \text{ J}$$
$$\text{Time taken} = 5 \times 60 \text{ s}$$
$$= 300 \text{ s}$$
$$\text{Power} = \frac{1500}{300}$$
$$= 5 \text{ W}$$

Compare this with a standard light bulb!

- *The quantitative links between kinetic energy, potential energy and work.* Kinetic energy is the energy a body has due to its speed. It is larger for greater speeds and for objects of higher mass since more energy has been put into these to give them such speed. Although the relationship is not an obvious one at first sight, pupils seem to find it easy to remember.

$$\text{kinetic energy} = \frac{1}{2} \times \text{m} \times \text{v}^2$$

The main problem with its use is that memorising *half mv²* encourages pupils to square *m* as well as *v*.

Example: Ed, of mass 50 kg, on a skateboard, travels at 10 m/s. What is his kinetic energy?

$$\text{kinetic energy} = \tfrac{1}{2} \times 50 \times 10^2$$
$$= \tfrac{1}{2} \times 50 \times 100$$
$$= \tfrac{1}{2} \times 5000$$
$$= 2500 \text{ J}$$

Potential energy can be due to strain or height since both situations are created by an energy input. Calculations are only required at this level for the gravitational potential energy.

$$\textbf{gravitational potential energy} = \textbf{weight} \times \textbf{height}$$
$$\text{J} = \text{N} \times \text{m}$$

You may find the expression written as $\mathbf{E} = \mathbf{mgh}$ since weight = mass × g (Chapter 4) and h = height raised.

How much potential energy is gained by a bird weighing 10 N flying from ground level to a treetop 20 m high?

$$PE = 10 \times 20 = 200 \text{ J}$$

KEY STAGE 4 ACTIVITIES

It is worthwhile to review the energy resources studied in Key Stage 3 and revisit the problems which cause environmentalists to regard the situation as a crisis. Ask pupils to look out for items in the press, since now even politicians see a cause for concern. A video is a good reviewing technique, allowing pupils to make up the questions themselves to challenge as a team provides opportunities for watching selected parts twice. A Scientific Eye video called simply *Energy* is very watchable, set in a context of arguing self-righteous teenagers, but will not be sufficiently challenging for more able Key Stage 4.

Set pupils the task of compiling a poster-sized table of the global problems referred to in the background, which will help to distinguish them and highlight their relative importance. Although not directly a response to the statement in the curriculum these effects are inevitably linked with the economical use of resources world-wide (see Chapter 14).

Efficiency

Measurements of efficiency in the school laboratory are fairly crude but suffice to underline which type of devices are more efficient. Clearly you need to be able to measure an energy input and an energy output. This rules out experiments using people to transfer energy, since the energy input is very difficult to measure. You can however use a range of simple machines, from a ramp to a pulley system.

The example above with the car jack illustrates the method of calculation. What must be measured for each machine is:

1 The load being lifted in newtons.

2 The distance it is being lifted *vertically* in metres.

The product of these two is the useful energy output.

3 The effort applied in newtons.

4 The distance the effort is moved in metres.

The product of these two is the energy input.

The above is an example using a mechanical input and output. For comparison pupils can find the efficiency of an electric kettle. The kettle's power output should be known (in watts on the serial number plate). Pupils will need to measure the time it takes to boil a known quantity of water. The calculation might seem a bit tricky but, once units are dealt with, it is still only the basic equation for energy.

$$\text{energy input (J)} = \text{power of kettle (W)} \times \text{time (s)}$$

From Chapter 14

$$\text{energy output} = \text{mass of water (kg)} \times 4200 \text{ (J/kg °C)} \times \text{temperature change (°C)}$$

Hence the efficiency can be found. Note that it is not necessary to weigh the water since 1 litre has a mass of 1 kg.

Generally the efficiency of electrical appliances is much higher than that of mechanical ones or those which burn fuels (don't forget, though, that the achievement of electrical energy at the power station involved an energy transfer in the first place).

Force and work

Pupils can find out the amounts of work they can do in some classroom activities such as step-ups, the arm (or cycle) ergometer, lifting sandbags, running up stairs etc.

You will need a set of bathroom scales reading in newtons. The sandbags should weigh 10 N or 20 N, if not tell pupils to weigh them. In each case if you are going on to calculate power pupils will need to time the events.

S 1 Step-ups – pupils can do a fixed number, say thirty, but will tend to be competitive so ensure that the step-up arrangement is very secure and can not tip up.

2 Ergometer – the cycle version would be rather expensive for most schools but the arm version forms a useful comparison with the step-ups. A rope passes over a wheel whose circumference is, conveniently, just 1 m. The rope is held in place by a spring–balance on one end and a weight, usually a kg, on the other. When you wind the handle clockwise you do work against friction. The force you apply is the difference between the hanging weight in newtons and the reading on the balance. The total distance moved is the circumference multiplied by the number of turns. (Note: the bearings can be damaged by turning the wheel the wrong way.)

3 Sandbags – these can be lifted on to the table. Give each group just two bags then as one is lifting, the partner timing can replace the first bag gently (so they don't encourage sand leaks) on the floor. Lift a fixed number.

4 Upstairs – the work done here will be the weight of the person multiplied by the *vertical* height of the stairs.

Power

In the above exercises, power can be calculated if the time for the activities has been measured. Simply divide the work done by the time taken in seconds. It is interesting to compare the activities for the arm muscles and the leg muscles. Typical answers are between 10 W and 20 W for the sandbag lifting and 100 W to 150 W for the stepping, reflecting the sizes of the relevant muscles.

KE and PE

Calculations of gravitational potential energy are carried out when considering work done in the above paragraphs. Kinetic energy is usually introduced through numerical problems to solve. The two are linked in the pendulum. Pupils will be able to find out the speed of the pendulum as it passes through the centre point of its swing. This is possible because the potential energy gained when the pendulum is lifted is converted to kinetic energy as it falls.

$$\text{PE at highest point} = \text{KE at lowest point}$$
$$\text{weight} \times \text{height} = \tfrac{1}{2} \times \text{mass} \times v^2$$

Note: The height is the vertical height through which the pendulum bob is lifted. From this relationship the speed of the pendulum bob can be calculated, though most pupils will need help rearranging such an equation. Interest will be increased if you can set up a really big pendulum (e.g. a broom suspended between two rails) and take 'bets' on the maximum speed of the pendulum.

The formula for kinetic energy can be applied to stopping distances dealt with in Chapter 6. If energy is to be conserved, the kinetic energy of a vehicle must be transferred before the vehicle can stop. Most will be transferred to the brakes as thermal energy. As this energy depends on the square of the velocity pupils can see the link with braking distances.

Assessing pupils' learning

- Assessment in this section will mainly be based on the solving of numerical problems. A range of examples will enable you to provide a challenge for any group, see Resources below.

As a guide:

EASY: Simple, numbers preferably under 10, no rearrangement of equations, quantities all given in the correct units.

MEDIUM: Larger numbers but no use of indices, or smaller (to one decimal place), can use triangle method, easy unit changes such as minutes to seconds.

DIFFICULT: Use of indices, decimal points. Should be confident with rearranging equations and converting any units as required. Providing 'new' situations.

- Research a simple machine and report, with demonstrations, on their benefits and disadvantages, to the group. They can be choose from a range of machines such as one of the many lever systems, a pulley system, gearing, a ramp, a screw (a rolled up ramp), a wheel and axle.

ENRICHMENT AND EXTENSION ACTIVITIES

Able pupils will be able to use the formula for energy stored in a stretched item such as an elastic band, a spring, bungey rope etc.

$$\textbf{potential (strain) energy} = \tfrac{1}{2} \times \textbf{force} \times \textbf{displacement}$$
$$= \tfrac{1}{2} \times \textbf{weight} \times \textbf{extension}$$

RESOURCES

Teacher resources

- There are lots of numerical questions to be found in most GCSE texts. You will find some suitable for a wide range of ability between: Johnson *GCSE Physics for You*, Century Hutchinson and Avison *The World of Physics*, Thomas Nelson.

6 Radioactivity

Nature of radioactivity

BACKGROUND

Radioactivity, together with astronomy, make up the only topics in the curriculum which include a majority of material largely developed in the last 100 years. In fact it was detected in 1896 by Henri Bequerel, after whom a unit of radioactivity is named. He noticed that a salt of uranium blackened a photographic plate by being near but not in contact with it. Some of the radiation could be stopped by placing thick metal, such as a key, on the plate. More active radioactive materials were isolated and investigated by Marie Curie, whose name is used for another unit. Years later she and other workers in the field developed cancers but in the early days the discovery was seen as having many medicinal uses.

The uses of radioactive isotopes frequently make news and pupils hear about the nuclear debate. For this reason it is a pity that its study is often left until the very end of the course and then rushed or skated over because of the difficulty of the conceptual models involved. I think that pupils could be introduced to the Geiger counter at Key Stage 3 and the use of powerful sources would be unnecessary (and indeed not allowed). They could gain an idea of background radiation and be shown the increased count when a piece of uranium-rich rock is present. The dramatic increase is remembered and engenders a healthy respect for this type of radiation.

However it is much more likely that all this work will be new at Key Stage 4. There are some pupils for whom such abstract concepts as isotopes and nuclear particles will be out of reach. Pupils need to know that the nucleus in the orbital model of the atom contains protons and neutrons. The number of protons determines the nature of the element, for instance, carbon atoms always have 6 protons in their nucleus. The neutrons can be seen as having a diluting influence on the strong repulsive forces between protons and the 'right' number must be present to result in a stable nucleus. Different numbers of neutrons can give rise to unstable nuclei, which will decay eventually. Note that neutrons do not hold a nucleus together against the electrostatic repulsion of the protons but merely reduce the force required to do so. The existence of other forces (called strong forces) has been suggested but that takes us beyond the realms of the National Curriculum to Key Stage 4.

S   **The safety implications are paramount and pupils must see that you take every precaution.** You must carry out all the investigations yourself (with an exception detailed below) as anyone aged under 16 may not handle radioactive sources. This includes never providing an opportunity for a source to go astray. You should find that the sources are kept in a locked cupboard that is not in a laboratory. You will sign them in and out. Once in the classroom you can never leave them unattended. Pupils should sit at least 2 metres away from any source. If a problem arises, such as a pupil being unwell, send for help or pack up the sources and take them with you.

Common misunderstandings
- All radiation is harmful and usually results in some sort of explosion.

KEY STAGE 4 CONCEPTS

Pupils should be taught:

- *That radioactivity arises from the breakdown of an unstable nucleus.* A model of a non-radioactive atom which suffices at this level is depicted as follows.

Name of particle	Position in atom	Charge	Mass
Proton	In central nucleus	+1	1 amu
Neutron	In central nucleus	0	1 amu
Electron	Orbiting at an great distance	−1	negligible

The unit is the amu, atomic mass unit.

Lighter elements have roughly equal numbers of protons and neutrons, higher elements have a greater number of neutrons in proportion. A nucleus of an element may vary as to the number of neutrons in the nucleus and these slightly different varieties are called isotopes.

A *nuclear* reaction is one that involves particles arriving at or leaving the nucleus. They must be distinguished from *chemical* reactions, which involve the orbiting electrons. An isotope may arise in two ways:

1 Naturally occurring; certain isotopes of elements are always present in fixed proportions and some of these are radioactive.

   Example: *Chlorine has two isotopes Cl^{35} and Cl^{37} which are present in the proportion 3:1. Neither isotope is radioactive. Carbon has two isotopes C^{12} and C^{14}, the latter in very small quantities. C^{14} is radioactive.*

2 Caused by the addition of a particle to a previously stable nucleus, this may take place in a particle accelerator to form isotopes for a range of functions (see Chapter 17). The added particle, usually a neutron, will be 'fired' from a particle accelerator.

   Example: *Radioactive iodine (I^{131}) can be use to treat an overactive thyroid gland. The radioisotope is selectively absorbed by the gland which is then gradually destroyed by the radiation the iodine emits.*

- *That there is background radioactivity.* Background radioactivity is present all over the Earth, all the time. The levels vary according to what particular sources may be present in the region.

Sources of radioactive radiation

In the air	Percentage of radiation received	Additional information
In the air	51%	
Ground and buildings	14%	Greater in Cornwall, Devon
Food and drink	12%	Through the food chain
Medical	12%	Higher percentage for patients using radiation therapies
Cosmic rays	10%	From space, higher with altitude
Miscellaneous	Less than 1%	Nuclear weapons testing, industrial outputs, occupational, nuclear power

Note: These statistics are published by the UKAEA and supplied by the National Radiological Protection Board.

Pupils have no need to memorise such statistics but should be aware that so called 'natural radiation' makes up most of what we receive in a year.

- *That there are three main types of radioactive emission, with different penetrating powers.* The three types of emission are called alpha (α), beta (β) and gamma (γ). The word emission is used here because the last is of a very different nature from the first two, namely it is a type of electromagnetic radiation whereas the first two are normally described as streams of particles at this stage.

Alpha particles have the smallest penetrating power. Investigation will show that it is only a few centimetres in air and is easily stopped by thin layers of solid material such as paper, foil or skin. It is, however, heavily ionising so can cause considerable damage within the small region it penetrates.

Beta particles are more penetrating, travelling 20–200 cm or more in air and requiring a thickness of several millimetres of aluminium to stop a reasonable percentage of them. A heavy metal such as lead is more effective at stopping radiation.

Gamma rays are very similar to X-rays and overlap with them in the electromagnetic spectrum. They are emitted with a range of energies whereas X-rays are man-made and usually have a narrow range of energies. Their range varies depending on the energy (frequency) but they can usually only be stopped by several centimetres of lead. The source of γ-rays available in school is stored in a lead container for this reason.

- *The nature of alpha and beta particles and of gamma radiation.* These properties as a whole can be found in most school texts but here is a summary for reference.

Alpha particles	Beta particles	Gamma radiation
Helium nuclei, 2 protons, 2 neutrons	Fast moving electrons	Electromagnetic radiation
Mass 4 amu	Mass very small	Zero mass
Charge +2 so effected by electric and magnetic fields	Charge −1 so affected by electric and magnetic fields	Zero charge so not affected by electric and magnetic fields
Heavily ionising	Weakly ionising	Negligible ionisation

Losing an alpha particle means a nucleus loses +2 of charge and 4 amu of mass

$$_{92}U^{238} \rightarrow {}_{90}Th^{234} + \alpha$$

Gamma radiation is given off at the same time.

Losing a beta particle means a nucleus loses no mass but loses -1 of charge, this has the same overall effect as gaining $+1$

$$_{90}\text{Th}^{234} \rightarrow {}_{91}\text{Pa}^{234} + \beta$$

Gamma radiation is given off at the same time.

Losing a gamma ray means no loss of charge or mass, only energy.

KEY STAGE 4 ACTIVITIES

Radiation

An opener for this topic could be the use of the Geiger counter. Show pupils the counter, a tube with a delicate end window, and connect it to the device you have for recording the counts or activity rate. Do not remove any safety covering from the window. Explain that it is able to count ionising radiation, meaning that the radiation creates ions by knocking electrons off air molecules. It is worth emphasising early on that we are not discussing radiation of other types such as light, microwaves etc. To distinguish, refer to the subject of this chapter as ionising radiation or, where the concept of ions would prove difficult, radioactivity radiation.

The counter records results in two ways, a numerical count and an audible click (look for a loudspeaker switch if you don't hear a click within 5 seconds). Just allow the class to listen.

- Clap when you think the next click is about to happen – this will lead to completely chaotic clapping so it underlines that you can not predict when radiation will enter the counter. It is a completely random process.
- *What is the count rate?* Allow pupils to measure the number of counts in 30 s, or this rate may be given by the counter.
- Just check this background count for a different 30 s period – it will vary because of the random nature. A common value is 1–2 counts per second (1–2 bequerels).
- *Where is this radiation coming from?* It is always present, though in some places it is greater than others depending on the sources that are around (see the list above). A proportion of it has actually travelled from outer space (cosmic radiation).
- Bring up a piece of uranium-containing rock, held in tongs, which should be in your radioactivity kit (see below) and amaze the class by the rapid increase in clicks as the rock gets nearer.
- *What other effects are caused by ionising radiation?* This raises the opportunity to show some demonstrations, according to the resources available.

Radioactivity kit

S The vast majority of schools have a kit, which contains their radioactive sources and the necessary tools along with them. Don't worry if the box appears to have contained more items at one time. This is because some sources, which were considered safe some years ago, are no longer used and have been disposed of. The container is signed out and in again to keep track of it use. **No pupil under the age of sixteen may use any of the sources.** When you use the sources always handle them with tongs, never put them near your eyes and only get one out of the box at a time. These precautions are mainly to establish their importance, they must be seen to be done. The container will bear the familiar hazard warning for radioactive materials. With solid sources there is no need for goggles or gloves. A typical kit will contain:

1 An alpha source – americium (Am^{241}).

2 A beta source – strontium (Sr^{90}).

3 A gamma source – cobalt (Co^{60}), in some older kits this may be quite weak now.

4 Radium (Ra^{226}) – emits all three types and is stored in a lead container.

5 A piece of radioactive rock.

6 Tongs and holders for the sources.

7 A board to slot the holders for the sources and the Geiger counter into.

8 A set of materials of varying thicknesses.

Some demonstrations of the effects of ionising radiation

It is hard to beat actually seeing the tracks made by alpha particles. These can safely be seen by pupils in a piece of apparatus called a Wilson cloud chamber. The source is enclosed within the chamber and, although it is necessary to remove the top to set it up, it is quite safe. Beneath the chamber is a compartment into which you put some dry ice, held in place with some foam. In the upper part you squirt about half a pipette of ethanol and then replace the lid. Staring down on to the top you will see tracks, which are straight lines, about 4 cm long, of droplets. These are drops of alcohol which have condensed on the ions created by the alpha particles. Only alpha particles are sufficiently ionising to make tracks in these simple cloud chambers. The appearance of the tracks is sometimes improved by rubbing the lid with a duster to create an electric field.

S   Discharging an electroscope is less impressive but shows the ionising properties of the radiation well. The electroscope can be charged by rubbing a polythene rod and transferring the charge from it to the plate of the electroscope. **Note:** do not remove the glass front, as you will probably end up spoiling the fragile gold leaf. When the leaf is standing out at an angle to the central rod bring up the radium source, directing it at the electroscope plate. You will see the leaf gently collapse against the central rod as the ions created discharge the electroscope.

S   A spark counter is quite lively. The device has a wire suspended above a piece of mesh. When connected to an EHT supply a potential difference exists between the wire and the mesh. Turn up the voltage cautiously until a spark is just ready to develop. You will probably turn it up too far but then just turn it down a little. However, avoid letting it spark more than absolutely necessary. The voltage will probably be in the order of 4000 V or 4 kV. This is quite safe if taken from the high resistance terminal (see Figure 16.1 of an EHT unit). Link terminal A to B before attaching to the black terminal on the counter. Connect terminal D to the red terminal on the counter. No link to terminal C. Hold the radium source over the wire and this will cause sparks to appear. They are like tiny lightning strikes jumping across the pathway made by the ions created by the radiation. Move the source around to avoid scarring one place. This is more effective in a darkened room.

The photographic effect of ionising radiation is important. It is the way in which radioactivity was discovered and is the basis of radiation workers' protection badges. Sometimes you can obtain a protection badge to show the class, if not a photograph of the photographic film will do. They are divided into three sections with different materials so that the type as well as the dose of radiation can be recorded.

The Geiger counter itself can be reintroduced now as a tool for discovering deposits of radioactive materials as well as for checking areas for contamination. Its operation is rather like the spark counter except that, instead of a spark, a small electrical impulse triggers the counter or loudspeaker (see any textbook for a diagram of a Geiger counter).

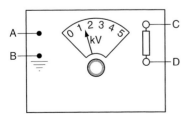

terminal A - black - negative
terminal B - black - earth
terminal C - red - positive
terminal D - red - high resistance positive

Figure 16.1

S **Experiments with the Geiger counter**

Important – check the Background comments and the section above on the Radioactivity Kit for safety procedures.

Alpha source

Set up the source at least 10 cm from the window of the Geiger counter. Move it slowly towards the counter. There should be a point at which the reading greatly increases. This shows that the particles have a fixed path in air from a particular source, as they all leave the nucleus with the same energy.

Place different materials between the source and the counter. Very thin layers will stop the particles from passing through. Pupils feel this makes alpha radiation safe. However such material ingested or breathed in causes a lot of damage because of the ionising effect within the cells.

Experiments with magnets are unlikely to show the particles being deflected as they are relatively heavy and would require very strong magnets.

Beta source

Set up the source at least half a metre away and proceed as before. This time the reading will increase more gradually as the low mass particles are easily knocked off course by air molecules they encounter.

Results for penetration are described above.

Deflection by magnets should show that beta particles are charged because the count rate reduces when the magnet is placed at the side of their path.

Gamma source

Investigate penetration in the same way as above but starting even further away. Pupils usually show great concern over the fact that their only protection from gamma rays is a substantial thickness of lead.

Since they are essentially the same as X-rays its worth pointing out that X-ray exposure should be kept to a minimum. The radiologist always retires to an adjoining room because he/she is exposed on a regular basis.

Presenting theory

Most of these investigations can be treated as interesting phenomena but some theory work must be included as appropriate for the group. It is important to find out where pupils are in their knowledge and understanding of particles. This might be done as a group activity giving each group a set of cards with statements on, to put into TRUE/FALSE/DON'T KNOW piles. According to previous experience their familiarity with the properties of different particles could reveal great variation but it is likely that a central nucleus model will have been adopted before Year 10.

It is necessary to get a feel for the nuclear model of the atom using coloured balls and a means of attaching them to one another. There are different ones available, negotiate with your Chemistry Department or glue some Velcro, criss-cross fashion like equator and polar orbits, on to large polystyrene balls. Use the model to show:

1 Alpha particles – stick together four larger balls, two coloured with pen or paint, forming a relatively massive particle. Show how it can be described as a helium nucleus and how it barges along knocking into other particles, ionising them and losing its energy quickly. It is not easily deflected by either electric or magnetic fields, because of its mass.

2 Beta particles – use some really tiny balls. Show how difficult it is for them to move in a straight path because they are continually bumping into bigger particles and being thrown off course. They are easily deflected by electric and magnetic fields.

3 Gamma rays – a laser beam would be useful here. Stress that it is still a model, show how it penetrates a tank of water, blocks of plastic, greaseproof paper etc. Gamma rays are even more energetic. They can not be deflected by either field (show this with a magnet at the side of the laser beam), and cause very little ionisation on their way through materials.

Assessing pupils' learning

- Write a book of advice for a hospital assistant who job is to prepare solutions of a radioactive substance for patients to drink.
- Draw a photographic record of the result of a radioactive meal! Discuss ideas first.
- Have a picture of a protection badge and describe its structure and reasons for it.

ENRICHMENT AND EXTENSION ACTIVITIES

Set a scene for a (relatively) minor nuclear accident, for instance a train crash. Pupils are the first to arrive on the scene – what do they do? Lots of imagination but correct science too.

More able pupils should be able to understand isotopes and may enjoy following a decay chain. Figure 16.2 shows an example.

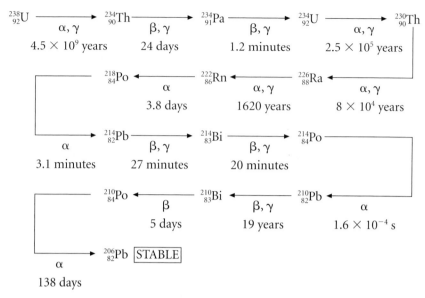

Figure 16.2 *The uranium 238 series.*

Pupils can research into chain reactions and their possibilities for good and harm.

RESOURCES

Video

- *Radioactivity* in the Physics in Action Series, Granada TV, is available to hire from The Institute of Physics, Education Department, 76 Portland Place, London W1N 3DH.
- UKAEA have a range of videos so it is worth having their catalogue. The relevant one here is called *All around us – Radiation*. Their address is UKAEA, 11 Charles II St, London SW1Y 4QP.

Pupil activity

- SATIS unit 807 *How much do you get?*

chapter 17

Benefits and dangers of radioactivity

BACKGROUND

Pupil research with plenty of resources provided is a useful way to approach this topic. As it is likely to be fairly late in the GCSE course major projects are not welcomed at this stage. In preference pupils can be given individual items to become expert in and bring them together to give an overall picture of the phenomenon that is radioactivity. This could be charted as an aid to revision because this is one section of the syllabus requiring quite a lot of memory work. There is, no doubt, a vast range of instances of the uses of radioactivity in industry, medical and research contexts but it will be easier for pupils to remember if the situations are ones that have some relevance to them. A few suggestions are given but with time you will gather more from keeping an eye on newspapers or science magazines and using the fruits of pupil research. Offer bonuses for anyone who brings in any news article concerning radioactivity.

Though not radioactivity, a topic on nuclear reactions is not complete without a look at nuclear fission and its contribution to the supply of electricity. For discussion of its relative merits many resources are available through the nuclear industry, but other points of view should also be represented, such as Friends of the Earth. They produce a leaflet on nuclear power which you can obtain, together with a list of their other publications, from Friends of the Earth, 26–28 Underwood Street, London N1 7JQ.

Common misunderstanding

• Two radioactive half-lives make one whole life.

KEY STAGE 4 CONCEPTS

Pupils should be taught:

- *The meaning of the term 'half-life'.* At any moment in time a sample of radioactive material has a certain activity (radioactive decays per second). With time, the activity falls because the number of atoms which have not decayed decreases. After a certain time, specific to each isotope, the activity has fallen to half. This time is called the half-life. After a second half-life the activity has fallen to a half again i.e. a quarter of the original, and so on. This half-life can not be changed by heating, pressure or chemical reaction and so can be used for time measurement purposes.

- *The beneficial and harmful effects of radiation on matter and living organisms.* When radioactivity was newly discovered it was thought to have many beneficial effects on living tissue from curing arthritis to improving the complexion. Packets containing tiny samples of radium and other active materials were strapped to the body to achieve these benefits. Of course, it is now known that such radiation plays a major part in causing cancer and also, in large doses, it causes sickness, burns and ultimately is fatal.

  Because of the ionising capabilities of the radiation, changes can be made in cells to the membrane, the nucleus, the chromosomes etc. These might be responsible for the cell dying or preventing division or they might bring about changes in behaviour. This in turn could lead to problems in individual organs such as the central nervous system, the intestines and bone marrow causing cancers. For an individual this could bring about early death. For gametes the changes could take place within the DNA which could then be responsible for changes in subsequent generations of the species. These changes could be beneficial but are much more likely to be harmful.

- *Some uses of radioactivity, including the radioactive dating of rocks.* As a tracer radioactive material is used to check for concentrations of elements or leaks of fluid from tubes.

  Example: I^{131}, *a beta source, can be injected into the blood stream to detect blood clots. Increased activity at a point locates the clot.*

  Tiny amounts of Co^{60}, a gamma source, can be incorporated in wires underground so their position can be readily detected.

  Thickness monitoring can be done by measuring the penetration through a substance or the backscatter from it.

  Example: *In a smoke alarm alpha particles are continuously collected at a steady rate. If smoke enters the space between source and detector the alpha particles do not reach the detector and the decrease in readings registers, triggering an alarm.*

Example: *Beta particles can monitor the thickness of a wide range of materials from tissue paper to sheet metal. Its activity, after passing through the material, is measured. Any change due to greater or less absorption, is used by computer to control the machinery maintaining the thickness.*

Cancerous cells can be destroyed by irradiation with gamma rays. The rays are focussed on to the diseased cells and the source manoeuvred so that contact with healthy cells is kept to a minimum.

Sterilisation of carriers of disease in animals helps to control outbreaks.

Sterilisation of instruments by killing the bacteria on them is in use and it is possible that similar techniques might be used to preserve food, should public opinion allow.

The energy given off by a radioactive isotope can be used to power a pacemaker or a remote motor.

Dating historical objects can be done making use of the known radioactive half-life of particular isotopes. The measurement of such very small activities is bound to have some uncertainty associated with it but nevertheless the technique has redefined the chronological order of many historical events and the time scale for the laying down of rocks. For living organic materials the percentage content of the C^{14} isotope is known. When the animal dies or the plant is cut down this proportion dwindles at a rate determined by the half-life of the carbon (5500 years). A comparison between the original activity and the current activity, in terms of the number of half-lives that have elapsed, reveals the age.

Example: *The Turin shroud was dated as originating between 1250 and 1350 AD.*

A constituent of rocks at a part of the rock cycle is radioactive potassium (K^{40} half-life is 1300 million years). This decays to a radioactive isotope of argon. If the activity of the argon can be measured then, assuming none at the formation of the rock, the age can be estimated.

Although not strictly part of the topic of radioactivity, nuclear fission must be included in a section on nuclear reactions. Figure 17.1 shows the radioactive material uranium 235 and the reaction which releases the energy for use in power stations.

When an unstable nucleus divides into fragments a vast amount of energy is released, together with neutrons. These can enter neighbouring nuclei and cause them to become unstable. They can then suffer fission in the same way, thus sustaining or increasing the rate of the reaction. This is referred to as a chain reaction. It can be controlled, as in a power station, or allowed to build up, as in a nuclear bomb. Most GCSE textbooks contain diagrams of nuclear power stations. When controlled, the energy is used to heat water or gas to

20

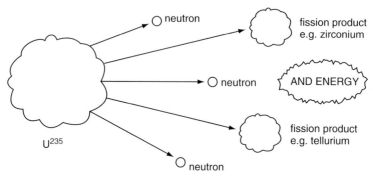

neutron

fission product
e.g. zirconium

neutron

AND ENERGY

fission product
e.g. tellurium

U^{235}

neutron

Figure 17.1

turn turbines (see Chapter 4). The problems associated with this source of power are:

1 Protection for the workers.

2 Monitoring to ensure that effluent is not radioactive.

3 Safety in operation – difficulties of controlling the chain reaction, by absorbing some of the neutrons, can mean that a power station can run out of control and possibly overheat or explode e.g. Chernobyl.

4 Disposal of spent fuel, which is still radioactive with a very long half-life.

5 Cost of decommissioning stations, which have come to the end of their useful life.

KEY STAGE 4 ACTIVITIES

Half-life

The concept of decay and the way in which it decreases with time can be modelled using large numbers and something that displays a random property. Usually dice are used. You really need at least 100 dice, preferably 100 for each group of 4 pupils. Provide them in boxes of 10 × 10 for easy retrieval.

Demonstrate first. Shake the dice in a plastic beaker and tip them onto a piece of paper. Remove and record the number of those that show a 6 (or any chosen number). The paper is to help you get them back into the container. Shake the remainder again and proceed as before until there are no dice left. Clearly as the number of dice goes down the number of 6s will go down. A plot of number of 6s against number of throws reveals a curve. As the process is random it may not be a very smooth curve so ask pupils how it might be improved. They may

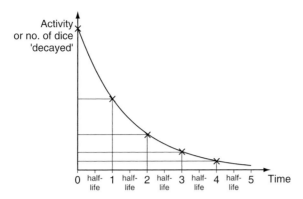

Figure 17.2

suggest using more dice. If lots of dice are available you can get groups to do the same activity then add all the results together to improve the curve; if not let just one set of volunteers have a go, to add to your own graph. The shape of the curve is described as exponential, and is characteristic of situations where the fall (or rise) of a variable is dependent on the number of the variable present (see Figure 17.2).

Radioactivity follows a similar decay curve but here the numbers are countless billions. Follow the diagram to explain how, no matter where you begin to count, the half-life is constant.

An element with a very short half-life provides a quick demonstration of the exponential curve. Protactinium is one that is particularly convenient since the isotope can be separated from the thorium which normally masks the activity, by dissolving in a solvent. The source comes already sealed in a container so all you have to do is shake the bottle, set it with the organic layer to the top and count the decay rate. Instructions are provide with the kit. Precautions and factors to improve reliability are:

1 Perform the whole experiment over a tray.

2 Wear gloves for handling the bottle.

3 Count and subtract the background count before introducing the source.

4 Do not touch the bottle to the Geiger counter.

5 Shake for 10–15 s and allow a few more seconds, after setting up, to settle.

6 Measure the activity for about 6 minutes and subtract the background count.

7 Plot a graph of activity in bequerels (Bq, counts per second) against time and average the several values of the half-life you extract from it.

Effects, benefits and harmful effects

As suggested above pupils would gain from researching a topic here and combining their results to make a giant chart. Give each pair a topic and get a report back to the class before building the chart. Specify the format for the presentation of their findings. In this way you will both ensure that they do the work and you will know what you have to accommodate on the chart.
Topics could include:

- half-life for dating historical documents, bones
- half-life for dating rocks
- radiation sickness
- cancers caused by radiation
- Chernobyl
- implications for offspring of irradiated organisms
- tracers in the blood, alimentary canal, lymph system
- tracers in pipes, cables
- tracers for testing engines, machinery etc.
- thickness monitors in smoke alarms, manufacture of any sheet materials, depth of powder in packets
- treatment of tumours, leukaemia
- sterilisation of harmful disease carriers and bacteria
- sterilisation of food

Some resources are listed overleaf.

Nuclear fission

Use the Velcro-polystyrene balls (see Chapter 16) to make a model for nuclear fission. Explain that fission only occurs in heavy atoms such as uranium but for simplicity you are using a smaller number of nuclear particles. Stick together at least nine, approach a neutron, at speed, and then wrench the nucleus apart into two unequal fragments plus a couple of neutrons. Show that each fragment has a smaller atomic mass than the heavier nucleus you began with and that the neutrons are free to latch on to other nuclei, repeating the effect. It is, unfortunately, impossible to show the energy being released at the same time. An animated video is best for this but I don't think it should replace the hands-on model of splitting.

The availability of this energy comes from the energy required to bind the nucleus together. The bigger an atom the more binding energy is required. If you imagine a nucleus being assembled from its protons and neutrons the mass of the nucleus is found to be slightly less than the totalled masses of the constituents. Einstein showed that energy has mass and he predicted that they were related by the famous equation $E = m \times c^2$ where E is the energy released, m is the mass difference and c is the speed of light. Given that the speed of light is 300 000 km/s it is not surprising that the quantities of energy are so huge. During fission the binding energy of the fragments is less than that of the heavy nucleus and so energy is released. If a kilogram of uranium were fully used in this way the quantity of energy released would be 80 million million joules; compare this with one kilogram of coal which releases 30 million joules when fully burned.

Assessing pupils' learning

- Provide a large clear diagram of a nuclear power station for pupils to complete, label and describe the functions, parts can be colour coded.
- Questions on half-life can be found in textbooks such as *GCSE Physics* by Tom Duncan.

ENRICHMENT AND EXTENSION ACTIVITIES

Students should debate the nuclear question leading to a vote on whether countries should be allowed to build more nuclear power stations. Use the SATIS units listed below. Note the UK is currently not considering building more, whereas France obtains 90% of its power from them and will continue with its programme of building.

Pupils familiar with spreadsheets can model radioactive decay. If a large number is chosen for the total number of atoms, say 1000, and a fraction for decay, say 1/10, then a formula, $1000 - 0.1 \times 1000$ can be entered as follows:

Unit of time	Number of atoms remaining
0	1000
(Enter formula to add 1 to cell above)	(Enter formula to subtract 0.1 × cell above from cell above)
Replicate down ▼	Replicate down until a small number remains ▼

Plot a graph of number remaining against time to obtain an exponential curve from which you can clearly see that half-life is constant.

RESOURCES

Video
- A range of videos is available from the UKAEA, (11 Charles II St, London SW1Y 4QP), including *Fast Reactor*, *Isotopes*, *Principles of Fission* and *Chernobyl – Could It Happen Here?*

Pupil activity
- Cartoon booklets on the reaction can be obtained free from Communications and External Relations, BNFL Magnox Generation, Berkeley Centre, Berkeley, Glos. GL13 9PB.
- SATIS Unit 204 *Using Radioactivity*.
- SATIS Unit 808 *Nuclear Fusion*.
- SATIS Unit 109 *Nuclear Power*.
- Spreadsheets include Microsoft Excel.

Teacher resources
- *GCSE Physics* by Tom Duncan is published by John Murray.

Examination questions

CHAPTER ONE

KS3

1 In this experiment the ball is moved towards the positive plate until they touch. It is immediately repelled towards the facing plate.

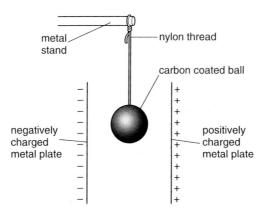

    metal stand — nylon thread — carbon coated ball — negatively charged metal plate — positively charged metal plate

a) i) Why is the ball repelled from the positive plate?

It has the same charge. Like charges repel.

1 mark

ii) Why is the ball attracted towards the negative plate?

It has the opposite charge. Unlike charges attract.

1 mark

b) What happens to the ball when it touches the negative plate?

It loses its positive charge and gains negative charge. It is repelled.

2 marks

c) Nylon is an insulator. Explain why an insulator is used to hold up the ball rather than a conductor like copper.

Otherwise the charge would be conducted away from the ball up the

thread.

1 mark

KS4

2 Figure 1 shows a model head wearing a wig. The model is earthed.

Figure 2 shows the same model and wig, connected to a + 100 000 V d.c. supply.

In each diagram the stand holding the head is made from an insulator.

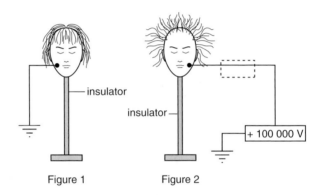

Figure 1   Figure 2

a) i) When charged by the + 100 000 V d.c. supply, what type of charge collects on the model?

A positive charge.

1 mark

ii) Draw an arrow in the dotted box on the supply line to show which way the electrons have moved.

→ The electrons all move to the right.

1 mark

iii) Explain why the hairs on the wig stand on end.

The hairs all have the same charge so they repel each other.

1 mark

b) Now a − 100 000 V d.c. supply is used instead of the + 100 000 V d.c. supply. What happens to the electrons and hair this time?

Electrons:

The electrons travel to the left.

Hair:

The hairs appear exactly the same.

(NEAB)
2 marks

CHAPTER TWO

KS3

1 The diagram shows a circuit for controlling an electric motor.

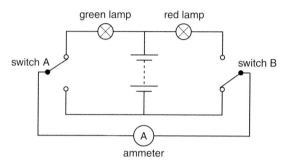

This circuit contains a centre-zero ammeter.

Complete the table to show which lamp, if any, is lit and in which direction, if any, the current flows.

The first row has been done for you.

Switch A	Switch B	Which lamp, if any is lit?	In which direction, if any, does the motor turn?
up	down	green lamp	left to right
up	up	*none*	*none*
down	up	*red lamp*	*right to left*
down	down	*none*	*none*

4 marks

KS4

2 The circuit diagram below shows a circuit used to supply electrical energy to the two headlights of a car.

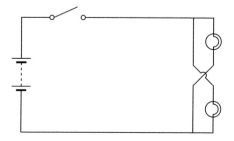

The current through the filament of one car headlight is 3.0 A. The potential difference across each of the two headlights is 12 V.

a) Suggest a suitable fuse for the circuit.

The fuse must allow at least 6 A. The nearest commonly available would be a 10 A fuse.

1 mark

b) Calculate the resistance of the headlight filament when in use.

$$resistance = \frac{voltage}{current}$$

$$= \frac{12}{3}$$

$$= 4\,\Omega$$

1 mark

c) Calculate the power supplied to the two headlights of the car.

$$power = voltage \times total\ current$$

$$= 12 \times 6$$

$$= 72\ W$$

2 marks

d) The fully charged car battery can deliver 72 kJ of energy at 12 V. How long can the battery keep the headlights fully on?

$$power = \frac{energy\ used}{time\ taken}$$

$$So \quad time = \frac{energy}{power}$$

$$= \frac{72\ 000}{72}$$

$$= 1000\ s$$

(NEAB)

2 marks

CHAPTER THREE

KS3

1 The diagram shows a bar magnet.

N

S

Explain each observation.

a) When floated in a dish on water the magnet always comes to rest in the same direction.

The magnet is influenced by the Earth's magnetic field. Its N-pole points to the Earth's N-pole.

3 marks

b) When placed near a nail the nail jumps towards the magnet.

The magnet induces the nail to become magnetised and attracts it.

1 mark

c) When placed near a block of steel the steel is pushed away.

The steel must be magnetised with the nearest pole facing the magnet, and so they repel each other.

1 mark

KS4

2 The diagram shows apparatus used to demonstrate the motor effect. *X* is a short length of bare copper wire resting on two other wires.

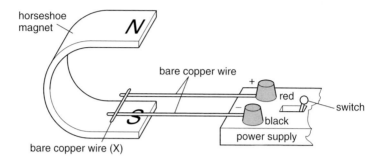

a) i) Describe what happens to wire X when the current is switched on.

The wire jumps or slides to the right (using the left hand rule).

ii) What difference do you notice if the following changes are made?

A The magnetic field is reversed.

The wire jumps to the left.

B The current is increased.

The wire moves more quickly since the force is stronger.

3 marks

b) The diagram shows a coil placed between the poles of a magnet. The arrows on the sides of the coil itself show the direction of the conventional current.

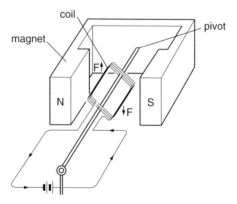

The arrows labelled *F* show the direction of the forces acting on the sides of the coil. Describe the motion of the coil until it comes to rest.

The coil rotates clockwise. When the plane of the coil reaches the vertical position there is no force acting but the momentum moves the coil a little further. The force is then reversed, turning the coil anticlockwise so it swings back. This cycle is repeated with the swing size being reduced

until it is at rest in the vertical position.

3 marks

c) Most electric motors use electromagnets instead of permanent magnets. State three of the features of an electromagnet which control the strength of the magnetic field obtained.

1   The size of the current.

2   The number of turns of the coil.

3   The iron core.

(NEAB)   *3 marks*

CHAPTER FOUR

KS3

1 Look at the diagram of the electric bell (Figure 4.1).

a) i)   Give the name of a suitable material used for the core of the electromagnet.

Iron (it must be magnetically soft).

1 mark

ii)   Give the name of a suitable material to use for the armature.

Iron (it must be magnetically soft).

1 mark

b) i)   Explain why the hammer hits the gong when the switch is closed.

When the switch is closed a current flows so an electromagnet is made OR

the iron core becomes magnetised. The armature is attracted to the core.

The hammer is attached to the armature.

3 marks

ii)   Why does the hammer repeat this action?

The contacts are broken so that the electromagnet stops working. When

the hammer is pulled back by the spring the whole cycle is repeated.

2 marks

KS4

2 a) Describe, in as much detail as you can, how the energy stored in coal is transferred into electrical energy in a power station.

Energy transfers from:

chemical energy in coal to thermal energy during combustion,

thermal energy to kinetic energy of steam,

kinetic energy to kinetic energy of turbine blades,

kinetic energy to electrical energy in wires.

5 marks

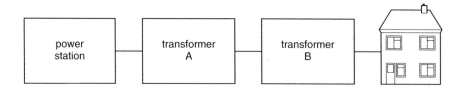

b) Transformer A produces a very high voltage to transmit the electrical energy through the National Grid.

Explain why electrical energy is transmitted at a very high voltage.

For a given power supply low voltage means high current. Power loss depends on the square of the current so current should be low, and hence voltage high.

3 marks

c) An appliance in a house has a transformer. The transformer is used to reduce the voltage to the level needed by the appliance.

The diagram below shows the transformer.

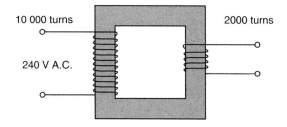

i) The transformer has 10 000 turns on the input side and 2000 turns on the output side. If the mains voltage of 240 volts is applied to the input, calculate the output voltage.

Note: You may find the following equation helpful:

$$\frac{\text{output voltage}}{\text{input voltage}} = \frac{\text{number of turns on output coil}}{\text{number of turns on input coil}}$$

$$\frac{output\ voltage}{input\ voltage} = \frac{2000}{10\ 000}$$

$$output\ voltage = \frac{2000 \times 240}{10\ 000}$$

$$= 48\ V.$$

ii) Name *two* factors which affect the amount of energy used by the appliance.

1   *The power of the appliance e.g. 100 W.*

2   *The time for which it is used.*

(NEAB)   *5 marks*

215

CHAPTER FIVE

KS3

1 A racer skates along a frozen river. Markers are set at intervals of 12 m along the bank. Once up to speed, he starts a stopwatch at a marker and stops it as he passes another marker 5 intervals later. The reading is 6 seconds.

a) i) What was his average speed?

$$speed = \frac{distance}{time}$$

$$= \frac{12 \times 5}{6}$$

$$= 10\ m/s$$

2 marks

ii) How could his speed over just one interval be found?

Measure his time between two adjacent markers.

1 mark

b) His top speed was found to be 12.5 m/s. How long did he take to go from one marker to the next at this speed?

$$time = \frac{distance}{speed}$$

$$= \frac{12}{12.5}$$

$$= 0.96 \text{ s}$$

2 marks

c) As the skater becomes more tired his speed changes. Would the time for each interval be less, more or the same as his average over the five intervals?

More – he takes longer because he travels more slowly.

1 mark

KS4

2 A racing driver is driving his car along a straight and level road as shown in the diagram below.

a) As the car's speed increases the forward force from the engine is bigger than the backwards force on the car. Eventually these forces become balanced. Explain why.

As the car gets faster the frictional force increases due to air resistance, moving parts, tyres etc.

3 marks

b Each lap of the circuit is 4500 m. The racing car takes 90 seconds to complete one lap. Calculate the average speed of the car.

$$speed = \frac{distance}{time}$$

$$= \frac{4500}{90}$$

$$= 50 \text{ m/s.}$$

4 marks

c) During the race the amount of fuel in the car gets less. Explain what effect this will have on the time taken to complete one lap.

The time will be reduced because the acceleration increases as the mass reduces for a given force. Force = mass × acceleration. Similarly deceleration will take less time.

3 marks

d) The racing car has a mass of 1250 kg. It is travelling at a speed of 48 m/s. When the brake pedal is pushed down a constant braking force of 10 000 N is exerted on the car.

i) Calculate the acceleration of the car.

$$acceleration = \frac{force}{mass}$$

$$= \frac{10\ 000}{1250}$$

$$= 8\ m/s^2$$

ii) The racing car travels a distance of 144 m before it stops.
   Calculate the work done in stopping the car.

work = force × distance

$$= 10\ 000 \times 144$$

$$= 1\ 440\ 000\ J.$$

(NEAB)

8 marks

CHAPTER SIX

KS3

1 a) This is a wagon standing still on a level track. Which two forces are acting on the wagon, and in which directions?

Gravity (or weight) downwards, and reaction upwards.

2 marks

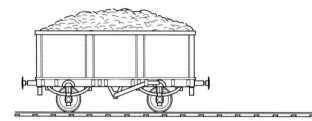

b) i) An engine is linked to the wagon and pulls to the right. Which way does friction act on the wagon?

To the left.

1 mark

ii) If the wagon does not move what does that imply about the size of the frictional force?

It is equal to the pull of the engine.

1 mark

iii) The engine pulls harder and moves the wagon forwards. Which way does the friction act of the wagon?

Still to the left.

1 mark

KS4

2 A car driver sees a dog on the road ahead and has to make an emergency stop.

The graph shows how the speed of the car changes with time after the driver first sees the dog.

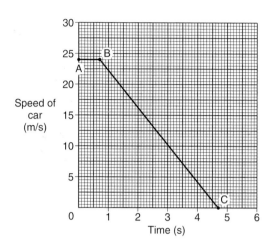

Speed of car (m/s) vs Time (s)

a) Which part of the graph represents reaction time or thinking time of the driver?

Section A-B because the care continues at constant speed.

1 mark

b) i) What is the thinking time of the driver?

0.7 seconds

ii) Calculate the distance travelled by the car in this thinking time.

distance = speed × time

$= 24 × 0.7$

$= 16.8\ m$

3 marks

c) Calculate the acceleration of the car after the brakes are applied.

$$acceleration = \frac{change\ in\ speed}{time}$$

$$= gradient\ of\ graph$$

$$= \frac{24}{4.7 - 0.7}$$

$$= \frac{24}{4}$$

$$= 6\ m/s^2$$

4 marks

d) Calculate the distance travelled by the car during braking.

The distance travelled is given by the area under the sloping line.

$$Distance = \frac{1}{2} \times 4 \times 24$$

$$= 48$$

$$= 48\ m$$

3 marks

e) The mass of the car is 800 kg. Calculate the braking force.

$$force = mass \times acceleration$$

$$= 800 \times 6$$

$$= 4800\ N$$

(NEAB) *3 marks*

CHAPTER SEVEN

KS3

1 A mechanic is trying to turn a nut with a spanner. He applies a force of 150 N at the distance shown in the diagram.

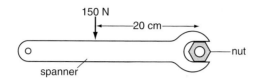

150 N

20 cm

nut

spanner

a) The spanner applies a moment to the nut. Calculate the size of this moment.

moment = force × distance from pivot

$$= 150 \times 20$$

$$= 3000 \, N \, cm$$

2 marks

b) What are the two changes he could make to increase this moment?

1 *He could increase the force he applies*

2 *He could apply the force further from the pivot.*

2 marks

KS4

2 A sky-diver steps out of an aeroplane.

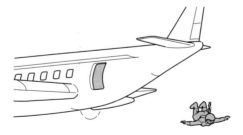

After 10 seconds she is falling at a steady speed of 50 m/s.

She then opens her parachute.

After another 5 seconds she is once again falling at a steady speed.

This speed is now only 10 m/s.

a) Calculate the sky-diver's average acceleration during the time from when she opens her parachute until she reaches her slower steady speed. (Show your working.)

$$acceleration = \frac{change\ in\ speed}{time\ taken}$$

change in speed after parachute opens = 50 − 10 = 40 m/s

time taken = 5 s

$$acceleration = \frac{40}{5} = 8\ m/s\ in\ the\ opposite\ direction\ to\ the\ motion$$

3 marks

b) Explain, as fully as you can:

i)   why the sky-diver eventually reaches a steady speed (with or without her parachute).

She accelerates until the air resistance force (upwards) is sufficient to balance her weight (downwards).

3 marks

ii)   why the sky-diver's steady speed is lower when her parachute is open.

Sufficient upwards force is achieved at a lower speed.

(NEAB)   *1 mark*

CHAPTER EIGHT

KS3

1 A catapult can be used to launch a plastic aeroplane. Energy is stored in the elastic band.

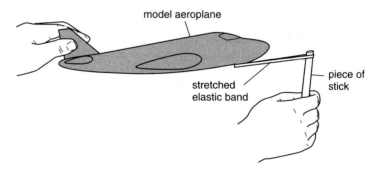

model aeroplane

stretched
elastic band

piece of
stick

a) How can *more* energy can be stored in the elastic band?

Stretch the elastic band more by pulling harder.

1 mark

b When the elastic band is released what energy transfer takes place?.

Potential strain energy is converted into the aeroplane's kinetic energy

1 mark

2 Some students have made a machine which will lift a 100 N weight using a smaller force.

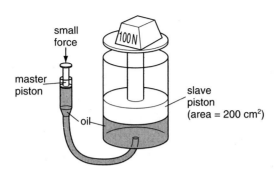

small force

master piston

slave piston (area = 200 cm²)

oil

a) Calculate the pressure of the slave piston on the oil. (Show your working.)

$$pressure = \frac{force}{area}$$

$$= \frac{100}{200}$$

$$= 0.5 \ N/cm.$$

3 marks

b) An area 2 cm² of the master piston is pressing against the oil.

How big a force must the students use on the master piston to lift the 100 N weight on the slave piston?

Explain your answer.

The area of the master is 1% of the loaded. Therefore the force is 1% of the load = 1 N

OR

$$Pressure \ on \ master = \frac{F}{2}$$

This equals the pressure on the slave since it must be the same throughout the liquid.

$$F = 2 \times 0.5$$

$$= 1 \ N$$

(NEAB) *2 marks*

CHAPTER NINE

KS3

1 Light shines onto a ball as seen in the diagram.

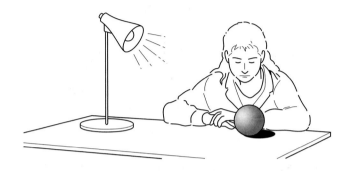

a) How can the student see the ball?

Light travels from the lamp to the ball. It is reflected from the ball and enters the eye.

2 marks

b i) Different coloured balls appear differently in coloured light.

Complete the table to show the colours that the balls appear.

colour of ball	colour of light	the colour the ball appears
white	red	*red*
green	white	*green*
green	red	*black*

3 marks

ii) Why does a black object appear black in any light?

A black object does not reflect any light falling on it OR a black object absorbs all/almost all the light falling on it.

1 mark

c) Choose from the following terms to complete the sentences below.

less than   equal to   greater than

At a plane mirror, the angle of incidence is *equal to* the angle of reflection.

The distance from the object to the mirror is *equal to* the apparent distance from the mirror to the image.

2 marks

2 a) The diagrams below show rays of light striking a mirror and a Perspex block.

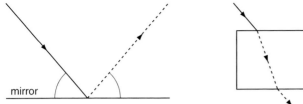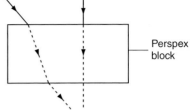

Complete the paths of the three rays of light on the diagrams to show the rays leaving the mirror and the Perspex block.

See the dotted lines.

4 marks

b) The diagram below shows a beam of light striking a Perspex block.

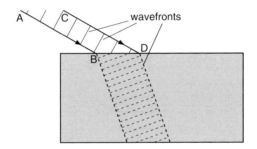

i) Continue the paths of the rays AB and CD inside the Perspex block.

See the dotted lines.

ii) Draw the wavefronts of the beam of light in the Perspex.

See the dotted lines.

iii) Explain why the beam behaves in the way you have shown.

Light travels more slowly in Perspex but the frequency is constant so the wavelength is smaller.

7 marks

c) The diagram below shows a ray of light striking a Perspex-air surface from inside the Perspex. The critical angle is 45°.

Draw the path of the ray after it reaches the Perspex-air boundary.

See the dotted line – the incident ray is at the critical angle

2 marks

d) The diagram below shows a fibrescope. This is an instrument which uses optical fibres to allow a doctor to see inside a patient's stomach.

Explain using the diagram if you wish, how the fibrescope works.

Light from the source enters the fibre at a large angle of incidence to the surface. Total internal reflection occurs, sending the beam to the stomach where some of it is reflected and travels in a similar way back to the doctor.

(NEAB)

6 marks

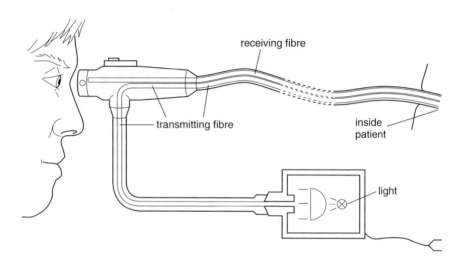

receiving fibre

transmitting fibre

inside patient

light

CHAPTER TEN

KS3

1 a) i) Hand-bell ringers create tunes using different sized bells. Smaller bells sound higher notes. What can you say about the frequencies of small and large bells?

Larger bells have a lower frequency and vice versa.

ii) Bells can ring loudly or quietly. What can you say about the amplitudes of the two notes?

The amplitude of the quiet ring is less than the amplitude of the loud ring.

2 marks

b) Which frequency is closest to that made by the bells?

2.5 Hz	☐
250 Hz	☑
25 000 Hz	☐
2 500 000 Hz	☐

1 mark

c) The energy given out when a bell rings was originally stored in the bell-ringer. Describe the useful energy transfers which take place when a bell-ringer rings a church bell.

Chemical energy in bell-ringer transferred to kinetic energy in the rope.

Kinetic energy in the rope transferred to potential energy in the bell.

Potential energy in the bell transferred to kinetic energy in the bell.

Kinetic energy in the bell transferred to sound.

3 marks

KS4

2 The diagram below shows a musician playing an electronic keyboard. The keyboard is plugged into an amplifier to produce sound. A microphone is placed in front of the loudspeaker so that the sound can be shown as a trace on an oscilloscope.

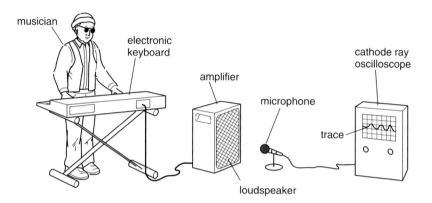

a) Explain how the sound travels from the loudspeaker to the microphone.

The loudspeaker cone vibrates and causes air particles nearby to vibrate.
The vibrations are passed along by a longitudinal wave to the microphone.

3 marks

b) The musician plays a note which has a frequency of 440 Hz. The speed of sound in air is 340 m/s.

Calculate the wavelength of the resulting sound wave.

$$wavelength = \frac{speed}{frequency}$$

$$= \frac{340}{440}$$

$$= 0.77\ m$$

4 marks

c) The diagram below shows the oscilloscope trace for a note of frequency 440 Hz

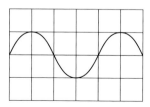

i) On the diagram below, draw in the trace you would expect to see if the same note was played, but louder.

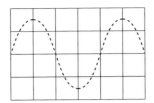

The wave has a larger amplitude

ii) On the diagram below, draw in the trace you would expect to see if a note of frequency 880 Hz was played.

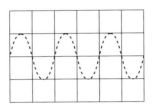

The frequency is doubled.

4 marks

d) Why is it not a good idea to listen to loud music, for a long time, using headphones?

It can cause ear damage which may lead to permanent deafness.

(NEAB)

2 marks

CHAPTER ELEVEN

KS3

1 a) When you drop a stone into a pond you hear the sound and see ripples spreading out across the pond.

Sound, ripples on water, and light travel at different speeds.

i) Which travels fastest: *sound* or *ripples on water* or *light*?

Light

1 mark

ii) Which travels most slowly: *sound* or *ripples on water* or *light*?

Ripples on water.

1 mark

b) Which sentence about sound is *correct*?

Tick the correct box.

Sound cannot travel through air. ☐

Sound cannot travel through stone. ☐

Sound cannot travel through a vacuum. ☑

Sound cannot travel through water. ☐

1 mark

KS4

2 Radio waves, ultra-violet, visible light and X-rays are all types of electromagnetic radiation.

a) Choose wavelengths from the list below to complete the table.

$3 \times 10^{-8}\,m$ $1 \times 10^{-11}\,m$ $5 \times 10^{-7}\,m$ $1500\,m$

Type of radiation	Wavelength (m)
Radio waves	*(1500)*
Ultra-violet	(3×10^{-8})
Visible light	(5×10^{-7})
X-rays	(1×10^{-11})

b) Microwaves are another type of electromagnetic radiation.

Calculate the frequency of microwaves of wavelength 3 cm.

(The velocity of electromagnetic waves is 3×10^8 m/s.)

$$frequency = \frac{speed}{wavelength}$$

$$= \frac{3 \times 10^8}{3 \times 10^{-2}}$$

(Remembering to change the units of wavelength into metres)

$$= 10^{10} \ Hz$$

4 marks

c) Which type of electromagnetic radiation is used:

i) to send information to and from satellites;

Microwaves

ii) in sunbeds:

Ultraviolet

ii) to kill harmful bacteria in foods?

Gamma rays

3 marks

d) Electromagnetic waves may be diffracted.

Explain the meaning of diffraction.

Diffraction occurs when waves pass through gaps. If the gaps are of a similar size to the wavelength, the waves appear to start out anew, spreading out as ripples from the gap.

(NEAB)

3 marks

CHAPTER TWELVE

KS3

1 The diagram shows planet Earth in orbit around the Sun.

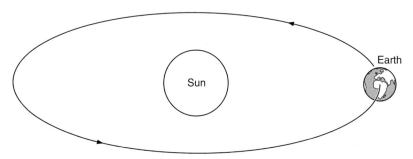

a) Two other planets are Mars and Venus. Which is closer to the Sun?

Venus

1 mark

b) Explain why Britain has both day and night every 24 hours.

Because Britain is spinning OR rotating every 24 hours [IT WOULD NOT BE ENOUGH TO SAY 'MOVING'].

2 marks

c) i) How long has it taken for the Earth to move to the position in this diagram from its position in the previous diagram?

9 months.

1 mark

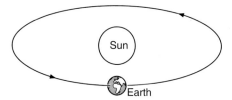

ii) Explain your answer.

The Earth completes one orbit in twelve months, so in nine months it will be three quarters of the way round.

1 mark

2 The diagram below shows the Solar System. It is not drawn to scale.

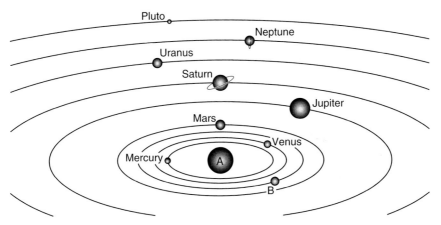

Pluto
Neptune
Uranus
Saturn
Jupiter
Mars
Venus
Mercury
A
B

a) The inner planets are very similar to each other, but they are also very different from the outer planets.

i) Write down *two* ways in which the inner planets are similar to each other.

1   *They are of a similar size.*

2   *They are of a similar density.*

ii) Write down *two* ways in which the outer planets are different from the inner planets.

1   *The outer planets are larger.*

2   *The outer planets are at a lower temperature [and/or density].*

4 marks

b) The diagram shows a satellite in orbit round the Earth.

i) Draw an arrow on the diagram to show the direction of the force which acts on the satellite.

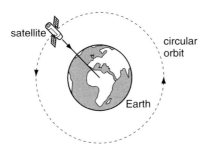

satellite
circular orbit
Earth

See diagram

ii) What causes this force?

Gravity

iii) Explain why the satellite is accelerating even though it is travelling at constant speed.

The satellite's direction is changing thus its velocity is changing. A changing velocity means an acceleration is taking place.

5 marks

c) Some satellites are placed in a geo-stationary orbit.

Explain what is meant by a geo-stationary orbit.

Its position does not change relative to the Earth.

1 mark

d) It is very expensive to launch satellites into Earth orbit.

State *four* uses of satellites which justify the very high launch costs.

1   *Communication*

2   *Mapping*

3   *Weather observation*

4   *Space exploration.*

4 marks

e)   Explain why:

i)   comets are not seen very often

They spend most of their time in outer space.

ii)   comets are only seen for a short time.

They are moving very fast and are only seen when reflecting the sun's light.

(NEAB)

4 marks

CHAPTER THIRTEEN

KS4

2 Astronomers use telescopes to observe stars and galaxies.

a) i) What is a galaxy?

A galaxy is a collection of stars numbering thousands of millions.

1 mark

ii) Briefly explain how stars are formed.

Particles of gases in regions called nebulae come together under the force of gravity. Whey they have gained sufficient mass fusion begins causing the star to emit radiation.

2 marks

b) The Hubble space telescope is in orbit round the Earth.

Astronomers can take better photographs through the Hubble telescope than by using telescopes on the ground.

Explain why.

There is no atmosphere or pollution to cause interference with incoming radiation from space.

(NEAB)   *2 marks*

CHAPTER FOURTEEN

KS3

1 A box falls from a height of 1 m to the ground. It loses 10 J of gravitational potential energy.

a) i) What happens to this energy while the box is falling?

It is converted to kinetic energy.

1 mark

ii) How much *kinetic* energy does the box have just before it hits the floor?

10 J (because the potential energy has been converted into kinetic energy).

1 mark

b) When the book hits the floor it stops and loses all its kinetic energy.

What happens to this energy?

The energy is transferred to the surroundings i.e. the air and the floor, raising the temperature by a very small amount.

1 mark

Energy comes from a variety of sources.

Complete the table opposite.

The first one has been done for you.

| | source of energy | | |
energy resource	directly from the Sun	indirectly from the Sun	not from the Sun
wind		✓	
nuclear			✓
hydro-electric		✓	
solar	✓		
geothermal			✓
oil		✓	

5 marks

KS4

2 A gas burner is used to heat some water in a pan.

Of the energy released by the burning gas by the time the water starts to boil

60% is transferred to the *water*.

20% is transferred to the *surrounding air*.

15% is transferred to the *pan*.

5% is transferred to the *gas burner* itself.

a) How does the energy which is transferred to the water compare with the energy transferred to the pan?

A quarter of the heat energy transferred to the water is transferred to the pan.

1 mark

b) Use the following information to explain the different amounts of energy transferred to the water and to the pan.

- The pan is made of aluminium.
- The mass of the pan and the mass of the water inside it are exactly equal.
- Both the pan and the water are heated from 20°C to 100°C.
- To raise the temperature of 1 kg of water by 1°C requires 4200 J of energy.
- To raise the temperature of 1 kg of aluminium by 1°C requires 900 J of energy.

Aluminium has a specific thermal capacity which is only about one quarter that of water i.e. it only needs one quarter as much energy to raise the temperature of the same mass by the same amount.

(NEAB) *3 marks*

KS3

A load is lifted by an electric motor.

1 The load weighs 2000 N.

   a) Calculate the work done in raising the load 20 m. Give the correct units.

$$work\ done = force \times vertical\ height$$
$$= 2000 \times 20$$
$$= 40\ 000\ J$$

2 marks

The load rises at a steady speed of 2 m/s.

   b) How long does it take to raise the load 20 m?

$$time\ taken = \frac{distance}{speed}$$
$$= \frac{20}{2} = 10\ s.$$

1 mark

   c) Calculate the power of the motor which would be needed to raise the load at a steady 2 m/s. Give the correct unit

$$power = \frac{work\ done}{time\ taken}$$
$$= \frac{40\ 000}{10}$$
$$= 4000\ W$$

2 marks

KS4

2 The diagram shows what happens to each 100 joules of energy from crude oil when it is used as petrol in a car.

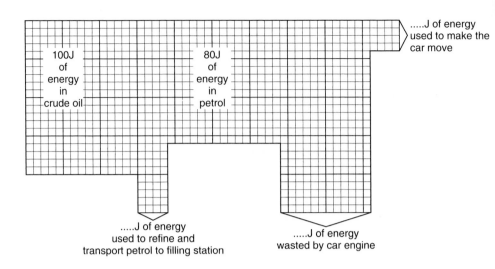

100J
of
energy
in
crude oil

80J
of
energy
in
petrol

.....J of energy
used to make the
car move

.....J of energy
used to refine and
transport petrol to filling station

.....J of energy
wasted by car engine

The widths of the arrows show exactly how much energy is transferred in each
particular way.

a) Complete the diagram by adding the correct energy value alongside each
arrow

In the diagram one large square represents 25 J of energy.

Movement = 20 J, waste = 60 J and refining = 20 J.

3 marks

b) Calculate how efficient the car engine is at transferring the energy *from petrol*
into useful movement. [Show your working].

$$efficiency = \frac{useful\ energy\ output}{energy\ input} \times 100\%$$

$$= \frac{20}{80} \times 100\%$$

$$= 25\%.$$

2 marks

c) Two students are discussing the diagram.

The first says that *none* of the energy released from the crude oil is really lost.

The other say that *all* of the energy released from the crude oil is really lost.

What do you think?

Explain your answer as fully as you can.

No energy is lost in the sense of destroyed but, because it is all eventually

dispersed, it is not available for use.

(NEAB)

4 marks

KS4

1 In an experiment to study a radioisotope a teacher used a special detector to measure the radioactivity. The detector produced an electric pulse when a radioactive particle entered it. The pulses were counted by an electrical counter.

a) Name a suitable radioactive detector.

Geiger counter.

1 mark

b) The teacher first used the detector to measure the background radiation level. This was done by switching it on for one minute. The background was found to be 24 counts/minute.

Suggest *two* sources of background radioactivity.

Any two of the following: cosmic radiation, radiation from building materials, sources in the area, naturally occurring isotopes such as carbon 14.

2 marks

c) The teacher then placed the radioisotope close to the detector as shown in the diagram below. The radiation reaching the detector in one minute was measured.

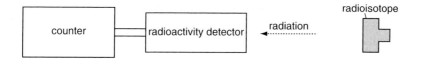

Then different materials were placed between the radioisotope and the detector. The radiation which reached the detector was measured each time. During the experiments the teacher took great care not to bring the radioisotope near the hands or to point it at anyone. The results of the experiments are shown in the table below.

Radiation Detected (counts/minute)

No material	Paper	Thin metal	Thick lead	Background count
895	460	24	24	24

i) What types of radiation did this radioisotope emit?

Alpha and beta particles.

1 mark

ii) Explain the reasons for your answer.

Alpha particles are absorbed by the paper while beta particles are absorbed by the thin metal. Gamma rays would have been reduced by the thick lead, so they can't have been present.

3 marks

d) Explain, as fully as you can, how the radioisotope could cause damage to your hand if you put your hand near to it.

Beta particles could enter the body through the skin. They create ions in their path. In the cells ions cause changes (mutations) which may harm or kill the cells or cause them to become cancerous.

4 marks

e) Radiation can also be used to treat cancer. A radioisotope that is often used is cobalt-60. This isotope has a proton number of 27 and a mass number of 60.

i) How many protons, neutrons and electrons does a cobalt-60 atom contain?

number of protons = *27*

number of neutrons = *33*

number of electrons = *27*

3 marks

ii) Radiation is used to treat cancerous growths. What does the radiation do?

If the radiation is directed at the cancerous cells and is sufficiently intense it will kill them.

(NEAB) *1 mark*

CHAPTER SEVENTEEN

KS4

1 Doctors sometimes need to know how much blood a patient has.

They can find out by using a radioactive solution.

After measuring how radioactive a small syringe-full of the solution is they inject it into the patient's blood.

They then wait for 30 minutes so that the solution has time to become completely mixed into the blood.

Finally, they take a syringe-full of blood and measure how radioactive it is

YOUR BLOOD CIRCULATION

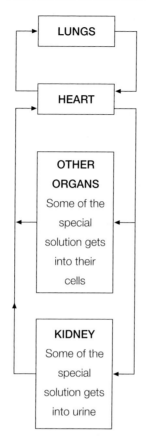

Example:

If the doctor injects 10 cm³ of the radioactive solution and this is diluted 500 times by the blood there must be 10 × 500 = 5000 cm³ of blood

a) After allowing for background radiation:

- 10 cm³ of the radioactive solution gives a reading of 7350 counts per minute;

- a 10 cm³ sample of blood gives a reading of 15 counts per minute. Calculate the volume of the patient's blood.
  (Show your working.)

The sample has been diluted by $\dfrac{7350}{15}$ times

So, the volume of blood is $\dfrac{7350}{15} \times 10\ cm^3$

$= 490 \times 10\ cm^3$

$= 4900\ cm^3.$

4 marks

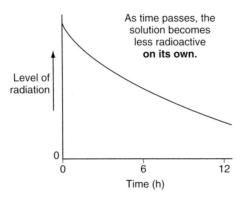

As time passes, the solution becomes less radioactive **on its own.**

Level of radiation

0

0   6   12

Time (h)

Radiation from radioactive substances can harm your body's cells

b) The doctor's method of estimating blood volume will not be completely accurate. Write down *three* reasons for this.

1   *In half an hour the count will have fallen away.*

2   *Some of the solution gets into the cells of other organs.*

3   *Some of the solution gets into the urine.*

3 marks

c) The doctors use a radioactive substance which loses half of its radioactivity every six hours. Explain why this is a suitable radioactive substance to use.

If the half life were longer there would be little noticeable difference after 30 minutes. If it were shorter the rate would fall too quickly in the half hour gap, making it very inaccurate.

(NEAB)   *2 marks*

Index

NOTES